Polymer Blends and Alloys

Polymer Blends and Alloys

Edited by

M.J. FOLKES
Brunel University
Uxbridge

and

P.S. HOPE
BP Chemicals Ltd
Grangemouth

SPRINGER-SCIENCE+BUSINESS MEDIA, B.V

First edition 1993
Reprinted 1995

Originally published by Chapman & Hall in 1993
Softcover reprint of the hardcover 1st edition 1993

Typeset in 10/12pt Times New Roman by
Thomson Press (India) Ltd., New Delhi

ISBN 978-94-010-4957-3

A catalogue record for this book is available from the British Library

Library of Congress Cataloging-in-Publication Data
Polymer blends and alloys / edited by P.S. Hope and M.J. Folkes.
p. cm.
Includes bibliographical references and index.
ISBN 978-94-010-4957-3 ISBN 978-94-011-2162-0 (eBook)
DOI 10.1007/978-94-011-2162-0
1. Polymers. I. Hope, P.S. II. Folkes, M.J.
TP1087.P643 1993
668.9-dc20 92-46269
CIP

Printed on acid-free paper, manufactured in accordance with ANSI/NISO Z39.48-1992 (Permanence of Paper).

Contents

Contributors

Dr P.T. Alder	Courtaulds Films, Cheney Manor, Swindon, Wiltshire SN2 2QF
Dr R.G.C. Arridge	H.H. Wills Physics Laboratory, University of Bristol, Royal Fort, Tyndall Avenue, Bristol BS8 1TL
Mr J.G. Bonner	BP Chemicals Limited, PO Box 21, Bo'ness Road, Grangemouth, Stirlingshire FK3 9XH
Mr C.S. Brown	National Physical Laboratory, Queen's Road, Teddington, Middlesex TW11 0LW
Mr J.E. Curry	Werner & Pfliederer Corporation, 663 E. Crescent Avenue, Ramsay, NJ 07446, USA
Dr M.J. Folkes	Department of Materials Technology, Brunel University, Uxbridge, Middlesex, UB8 3PH
Dr P.S. Hope	BP Chemicals Limited, PO Box 21, Bo'ness Road, Grangemouth, Stirlingshire FK3 9XH
Dr W.H. Lee	Department of Materials Technology, Brunel University, Uxbridge, Middlesex, UB8 3PH
Professor J. Lyngaae-Jørgensen	Technical University of Denmark, Instittet for Kemiindustri, DTH Building 227, DK-2800 Lyngby, Denmark
Dr S.C. Steadman	PERA, Melton Mowbray, Leicestershire LE13 0PB
Dr D. Vesely	Department of Materials Technology, Brunel University, Uxbridge, Middlesex, UB8 3PH

1 Introduction

P.S. HOPE and M.J. FOLKES

Mixing two or more polymers together to produce blends or alloys is a well-established strategy for achieving a specified portfolio of physical properties, without the need to synthesise specialised polymer systems. The subject is vast and has been the focus of much work, both theoretical and experimental. Much of the earlier work in this field was necessarily empirical and many of the blends produced were of academic rather than commercial interest.

The manner in which two (or more) polymers are compounded together is of vital importance in controlling the properties of blends. Moreover, particularly through detailed rheological studies, it is becoming apparent that processing can provide a wide range of blend microstructures. In an extreme, this is exemplified by the *in situ* formation of fibres resulting from the imposition of predetermined flow fields on blends, when in the solution or melt state. The microstructures produced in this case transform the blend into a true fibre composite; this parallels earlier work on the deformation of metal alloys. This type of processing–structure–property correlation opens up many new possibilities for innovative applications; for example, the production of stiff fibre composites and blends having anisotropic transport properties, such as novel membranes.

This book serves a dual purpose. On the one hand it provides an up-date on the more conventional activities in the field of polymer blends, with a focus on practical aspects of use to the technologist interested in the development of blends or the design of production processes. To this end chapters are included on the rheology, processing and compatibilisation of blends, along with practical introductions to techniques for studying microstructure, and for modelling and understanding their mechanical performance. Even more important, it will introduce workers in both industry and academia to some of the forward-looking possibilities which the informed processing of polymer blends can offer, including the use of liquid crystal polymers and the novel development of *in situ* fibre composites.

Starting with the important area of blend production, Curry, in chapter 2, addresses some practical aspects of polymer blend processing, concentrating mainly on describing the operating principles and performance characteristics of compounding equipment suitable for the preparation of blends and alloys. The main features of the many different commercially available compounders are presented, with reference to their historical development, and practical studies of the mixing mechanisms which operate in such processes are reviewed. Consideration is also given to the fundamental mechanisms of polymer dispersion, which are primarily responsible for the final phase morphology of immiscible blends. Theoretical and model experimental approaches to the issues of droplet break-up and coalescence are discussed, and it is concluded that these fundamental studies currently cannot predict the outcome of commercial processes, far less be used to design them; however, they do lead to some general rules which are useful for the technologist involved in blend development. The formulation of these guidelines is followed by a study of the performance of five commonly used compounders, carried out on both immiscible blends of low density polyethylene and polystyrene, and similar blends in the presence of a compatibiliser. It is clear from this work that choice of compounder and operating conditions affects the phase morphology of the resulting blend or alloy, and moreover that simply increasing the specific mechanical energy input to the compounding processes does not guarantee production of finer morphologies.

Compatibilisation, that is to say modification of normally immiscible blends to give alloys with improved end-use performance, is an important factor in almost all commercial blends, and has been the subject of an enormous amount of experimental investigation, much of which remains proprietary. The reasons for compatibilising blends, and the strategies which can be employed, are dealt with in some detail by Bonner and Hope in chapter 3. The main methods of achieving compatibility are via addition of block and graft copolymers, or functional/reactive polymers, and *in situ* grafting or polymerisation (reactive blending). Which approach to employ will depend very specifically on the blend system under consideration, the technology available and (not least) the cost. For the convenience of the user the literature is reviewed in two parts; blend systems where separate compatibilisers are added, and those where compatibility is achieved by using reactive blending technology. Particular attention is paid to the most extensively researched systems (polyethylene/polystyrene blends, and blends containing polyamides). Future trends in which compatibilisation holds the key to commercial success are also identified. These include engineering polymer blends, aimed at very high performance applications such as aerospace products, superior performance commodity polymers, in which improved performance (e.g. thermal or permeation resistance) are imparted by addition of engineering polymers, and polymer recycling, for which compatibilisation may offer a more cost-effective alternative to separation technology.

In chapter 4 the rheology of polymer blends is explored by Lyngaae-Jørgensen, with particular regard to its role in the development of phase morphology during the production and processing of immiscible blends. The main rheological functions are introduced, followed by a summary of the rheological behaviour of miscible blends, blends exhibiting phase transitions during flow and immiscible blends. For immiscible blends in particular it is clear that the rheology can be extremely complex, and that although recognisable and useful rheometric data can be obtained, unusual relationships, e.g. between viscosity and composition, are commonplace. Interpretation of these complex functions demands a 'microrheological' approach, taking into account the phase structure and the interfacial characteristics of the component polymers. Theoretical approaches to the issues of droplet break-up and coalescence during flow, introduced in chapter 2, are presented in more detail, including extension to include non-steady-state flow, and are illustrated by a case study for structure development during simple shear flow.

Chapter 4 also addresses the more complex flows which occur in extrusion and moulding processes. Experimental studies of extrusion and batch mixing processes have shown that most of the significant morphology development occurs very early in the process, mainly in the melting and softening stages. Prediction of the behaviour of polymer blends in all stages of an extrusion process—typically solids feeding and flow, melting, mixing, melt flow and solidification—is the subject of much current research but a total simulation is unlikely to be achieved in the foreseeable future, if ever. For this reason a more pragmatic approach based on dimensional analysis is advocated, which revolves around characterising the process in terms of dimensionless droplet break-up time, component viscosity ratio and capillarity number. Using this approach it is possible to identify some practical and quantitative guidelines for predicting the development of blend morphology during compounding processes.

As well as influencing—and being influenced by—rheology, as discussed above, the microstructure of polymer blends, and in particular their phase morphology, plays a major role in determining end-use performance. It is therefore important for those who are developing new blends, and indeed for technologists working with existing blends, to have available methods for studying phase structure and other microstructural features. In chapter 5, Vesely provides a practical guide to techniques for studying blend microstructure. The focus is mainly on microscopy, which in its various forms (light microscopy, scanning and transmission electron microscopy) provides the most useful and direct tool for observation of phase morphology. Details of operating principles and techniques are described, and practical advice is given on the application of a range of microscopy methods to polymer blends. Thermal analysis methods are also extensively used, particularly for measuring glass transition temperatures and characterising melting/crystallisation behaviour, and their application to blends is outlined. Other techniques (light scattering, neutron and X-ray scattering, and spectroscopy) are also briefly described. It is clear that in many cases combinations of measurement techniques are required to get a sufficiently useful picture of microstructure.

Chapters 6 and 7 are concerned with the end-use performance of polymer blends. Arridge (chapter 6) tackles theoretical aspects of polymer blends and alloys and explains, in some mathematical detail, the theory which exists for predicting the mechanical properties of composites and blends from component properties. For real blends and alloys this rigorous approach is essentially limited to the prediction of small strain behaviour (e.g. elastic constants) using linear viscoelastic representations. After general introductions to continuum mechanics, viscoelasticity and methods of representing structural anisotropy, predictive equations applicable to blends and alloys are presented and limitations to their use are discussed. The approach taken to predicting the properties of inhomogeneous (phase separated) blends is to treat them as composite materials, for which the properties of the composite phases can be combined using mixing rules of varying complexity. This provides a sound basis for understanding mechanical performance and will serve as a guide to further study for those involved in developing new materials.

Toughened polymers represent a large and rapidly evolving application area for polymer blends, particularly for engineering resins, where the requirements in terms of specific mechanical performance, sometimes at light product weights and at high operating temperatures, are exceptionally demanding. Established high-performance systems are repeatedly modified and improved largely by innovations in synthesis and compounding to produce more effective blends and alloys. In chapter 7 Lee provides a comprehensive survey of the toughening methods available for use in thermoplastics and thermosets. For thermoplastics, where toughening is achieved predominantly by incorporation of rubbery inclusions, the compatibilisation methods introduced in chapter 3 are revisited, but with emphasis on the more recent advances related to toughened polymers. More attention is paid in this chapter to toughening methods for the novel thermoset resins, including epoxy, cyanate ester and bismaleimide resins, developed in recent years as matrices for fibre-reinforced aerospace composites. The different processing technology associated with thermosets demands different toughening strategies. These include not only the introduction of rubbery inclusions but also incorporation of thermoplastic or rigid particulate fillers, and the use of semi-interpenetrating networks. The diverse mechanisms by which toughening can be achieved are reviewed, and related to specific material systems. These include shear yielding, crazing and voiding of the matrix polymer, cavitation, tearing and ductile drawing of rubber particles and other soft inclusions, and crack-pinning mechanisms. The role of particle size distribution and the influence of processing on toughness are also discussed.

Chapters 8 and 9 describe two of the new developments in polymer blends which offer potentially exciting future opportunities for commercial materials. Brown and Alder (chapter 8) present a review of work to date on the development of blends containing liquid crystal polymers (LCPs). The unusual flow and subsequent mechanical property behaviour of LCPs, which under suitable conditions can exist as structurally ordered fluids, makes them interesting materials in their own right. Commercial interest is growing in their use in blends, although their high cost currently inhibits exploitation. One significant feature of blends containing thermotropic LCPs is their ability in some cases to significantly alter the rheology of the blend, often giving reductions in viscosity. Although the mechanisms by which LCP addition modifies flow behaviour are not well understood, a variety of effects have been observed in processing blends containing LCPs, some of which may prove to be of practical interest (for example reductions in melt temperature, melt pressure and extruder torque). A second feature of LCP blends is their ability to form a fibrillar structure during processing, giving a self-reinforcing effect, although poor interface adhesion and difficulty in controlling phase morphology have so far prevented realisation of potential. However, the study of blends containing LCPs is in its infancy and could offer exciting future possibilities.

The field of fibre-forming blends, and *in situ* composites, is itself an emerging technology, ultimately offering the potential to manufacture high performance composites via less expensive production routes than those currently used. In chapter 9 Steadman describes these complex systems, in which fibre-forming polymers are introduced to host matrices and then oriented to give reinforced structures. Methods of achieving this include *in situ* crystallisation or polymerisation of the fibre-forming polymer and the use of thermotropic LCP blends. Clearly the technology depends for its success on understanding numerous factors; for example how to control and achieve the desired blend morphology, including the effective use of compatibilisers, the response of the fibre-forming phase to elongational (orienting) flows during forming and the ability to control the placement of orientation-inducing flows in complex geometries. After a review of the field, a detailed study is presented using high density polyethylene as the fibre-forming phase and a polystyrene–polybutadiene–polystyrene (SBS) block copolymer as the matrix. Blends produced on a co-rotating twin-screw extruder and subsequently ram extruded or uniaxially drawn at a range of temperatures produced composites containing polyethylene fibrils with stiffnesses equivalent to E-glass fibres. This highly promising result points the way to an exciting future if the technology can be harnessed for use with other thermoplastic matrices.

Finally, it is clear that a book of this nature cannot comprehensively cover a field which is enormous in terms of the breadth of science and technology and the diversity of materials systems employed, and which is growing rapidly. It does, however, pull together the key technology issues in sufficient depth to provide the newcomer with a grasp of their scope and significance and to point the way forward in some of the emerging new technologies. This will no doubt prompt the reader to enquire further in his or her area of interest; to assist in this the bibliography below contains a few of the excellent books published in the field in recent years. Utracki (1989) in particular contains recent and comprehensive summaries of patent literature and commercially available blends. However, any amount of reading can do little more than provide the tools needed for the job; the way forward in this fertile area must then be left to the ingenuity of the blend developer.

Bibliography

Paul, D.R. and Newman, S. (eds) (1978) *Polymer Blends*, Academic Press, New York.

Utracki, L.A. (1989) *Polymer Alloys and Blends: Thermodynamics and Rheology*, Hanser, Munich, Vienna, New York.

Walsh, D.J., Higgins, J.S. and Maconnachie, A. (eds) (1985) *Polymer Blends and Mixtures*, Martinus Nijhoff, Dordrecht.

2 Practical aspects of processing of blends

J.E. CURRY

2.1 Introduction

This chapter appears amid an environment of intensive investigation into the thermodynamics and flow behavior of polymer blends themselves and an enhanced understanding of the kinematic behavior of mixing devices. As in other fields, the digital computer has multiplied the detail with which physicists can explain the observed development of blend morphology and engineers can elucidate on the deformation and stress effects of their mixer design. An argument could be made to let the controversy settle before going to print. On the other hand, a consensus of identifying cause and effect has developed which should be published.

The objective of this chapter is to present the machinery, techniques and analysis of commercial polymer blending processes.

2.1.1 History

The development of modern compounders with their complicated gear trains, precision tooling, high tech metallurgy, and intricate process geometries can be traced [1, 2] through a history of personal rivalries, international cooperations, corporate cartels, military urgency and war reparations. What has evolved represents economic and marketing successes as well as reliable technical solutions.

The earliest compounders were developed as dough mixers in the food industry and coincidentally as mixers for viscous chemicals like rubber, gutta percha, masonry products, and soap. Mixers and extruders were originally developed to separately mix and then transport or shape viscous masses.

2.1.2 Mixers

Single-shaft (Figure 2.1). Plodders, paddle mixers, and ribbon blenders are used primarily for powder blending and, sometimes, in the jargon of plasticized polymers, for preparation of the 'dry mix', a plasticizer loaded polymer powder. The mixers may have close fitting rotors, heated barrel jackets, alternately pitched rotors and barrel pins to enhance mixing, as well as

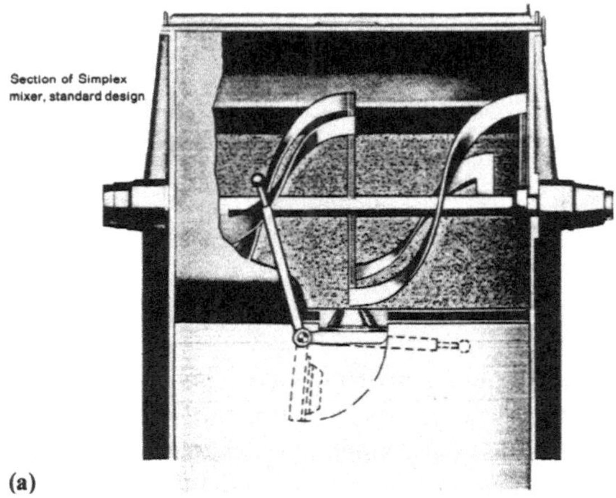

(a)

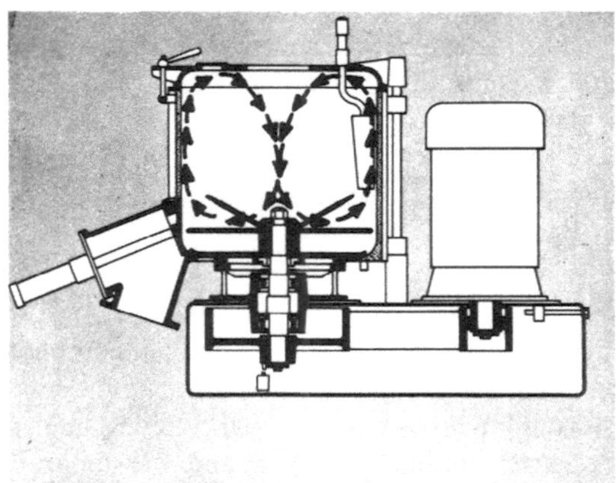

(b)

Figure 2.1 Single-shaft mixers used for feedstock blending. (a) Low intensity ribbon blender, simplex model (courtesy of Werner & Pfleiderer Corporation). (b) High intensity blender model MA (courtesy of T.K. Fielder).

ingenious dump and opening mechanisms to allow complete emptying and access for cleaning. These mixers are not normally used for plasticating processes.

A second class of single-shaft mixers is the intensity blender which is distinguished by its rugged design and high shaft speed with tip speeds in excess of 40 m s^{-1}. In addition to mixing, subtle morphology changes can be effected by the so-called thermokinetic action. These changes include blend sintering and powder densification.

Multishaft. Deficiencies of single-shaft mixers include an inability to generate high mechanical stress fields throughout a large portion of the mix volume and a tendency of plasticated mixes to stick to the rotor.

From the 1840s onward, generations of mills, cutters and multishaft mixers were invented in the industrial centers of the time; the US, England, and industrialized states of current Germany. These devices melted, mixed, stuffed and coated rubber, doughs, sausages, ceramics and the other viscous masses of the day on a commodity scale for the first time.

In the early 20th century a highly successful class of internal mixers was developed which survives and has spawned a continuously operating counterpart. In 1916, F.H. Banbury of Werner & Pfleiderer (Saginaw, Mich.) was granted a patent for a counter-rotating batch mixer whose shafts rotate at different speeds and whose charge is contained by a ram (Figure 2.2). The rotor design which caused flows to circulate through the chamber was an adaptation

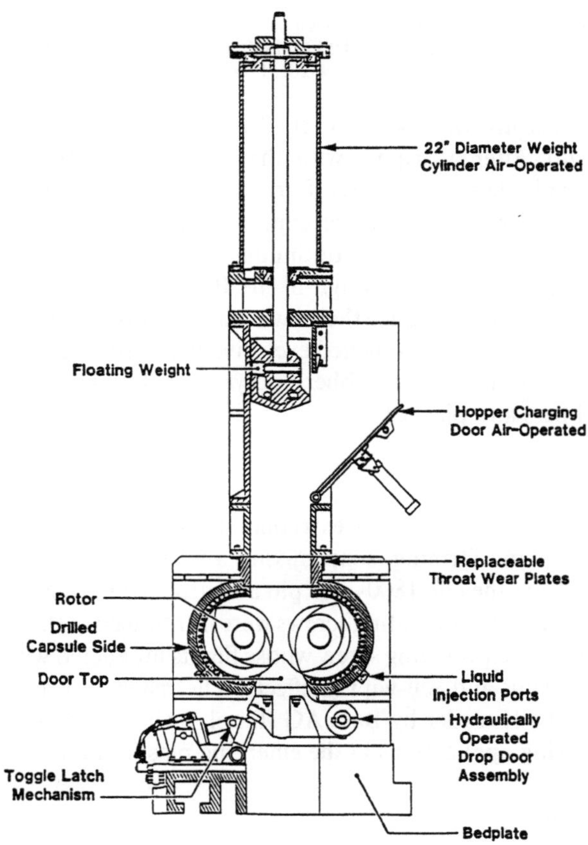

Figure 2.2 Plasticating multi-shaft batch mixer after the Banbury Patent, Model F270 (courtesy of Farrel Corporation).

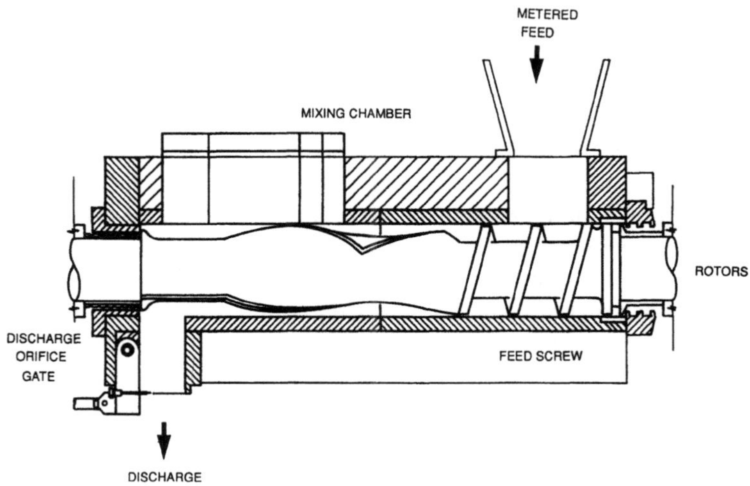

Figure 2.3 Adaptation of rotor type batch mixer as a continuous compounder. Counter-rotating non-intermeshing plasticating mixer. Model FCM (after P. Hold, *Advances in Polymer Technology*, **4**, 281, 1984).

of previous designs of engineers at Werner & Pfleiderer and Perkins (Peterborough). After a dispute with the management, Banbury took his design to Birmingham Iron (later Farrel Birmingham, Ansonia, Ct). The Farrel mixer dominated the rubber compounding business in the following years. Farrel later developed a continuous compounder, the FCM, which maintained the winged rotor design of the Banbury (Figure 2.3). Werner & Pfleiderer continued to produce their own winged rotor mixer, the GK mixer. While the GK mixer was channelled exclusively into rubber applications, the Banbury was also applied to rubbery polymer melts like polyethylene and blends of rubbers and plastics.

2.1.3 Extruders

Single-screw. Plasticating screw extruders started appearing in the 1870s for rubber and ceramic forming, soap mixing and wire coating. Screw design variants date from the late 1800s and pin and grooved barrels followed in the early 20th century. Reciprocating screws were introduced in the 1940s. In 1947 List developed a reciprocating screw whose flights interacted with a pin barrel to produce mixing and self-wiping effects. The machine was commercially developed as the Kneader by Buss AG (Basel, Switzerland) and was eagerly applied to mixing applications in the emerging plastics industry.

Counter-rotating

(i) Non-intermeshing. In a separate legacy to the FCM development, Welding Engineers (Norristown, PA) concentrated on the development of

counter-rotating machines for the handling of clay and catalyst in the pre World War II years. After the war, Fuller was issued a patent for the design of a twin-screw counter-rotating, non-intermeshing extruder. The screws turned together at the top, an innovation that was exploited in the devolatilization processes. The original design was soon enhanced with modular screw elements which effected transport and mixing tasks. Screw flights of matched and staggered arrangement on each shaft are known. Successes in synthetic rubber degassing were followed by developments in mechanical filtering of raw latex. In the 1970s, Welding Engineers capitalized on their capability to build long machines and combined dewatering, degassing and thermoplastic addition into a single process for the manufacture of rubber toughened blends.

(ii) Intermeshing. Fully intermeshing counter-rotating screw pumps have been known since the early 1870s for pumping doughs, oils and aqueous slurries. In 1935, the Leistritz company was granted a patent for a kneading counter-rotating pump which they developed in partnership with I.G. Farben. A kneading effect was developed by decreasing the pitch of the screw so that the pump chambers would overflow. Originally applied to oil, tars and ceramics, the machine was also applied to I.G. Farben's polymer products. The Leistritz plant was destroyed in World War II and the machine resurfaced as a polymer compounder in the 1950s.

Co-rotating intermeshing. Co-rotating intermeshing designs were developed and applied in the 1930s for mixing and pumping of ceramic pastes. The first commercially produced machine was invented by Colombo of LMP (Turin) in 1939, and used for thermoset and PVC processing.

In a separate development Erdmenger and colleagues at I.G. Farben developed co-rotating machines for processing of plastic masses. In 1949 Erdmenger (with Bayer) developed co-rotating extruders with modular elements and kneading discs for improved mixing. In 1953, Bayer licensed Werner & Pfleiderer (Stuttgart) who expanded the Erdmenger development to a commercially produced series of machines. Bayer and the other engineering giants of the German chemical industry provided a stream of applications in polymerization, devolatilization, plastification and mixing of thermoplastics.

In yet a third independent development, Loomans and Brennan of Readco developed an intermeshing, self-wiping extruder with screws and paddles. The patent was transferred to Baker Perkins who developed commercial equipment originally applied to condensation reactions.

Specialized devices. Screw devices are the prevalent but not exclusive machines for plasticating and compounding. Batch devices have the advantage of arbitrary residence time according to process needs. Several firms have developed cascaded batch-like systems which can be configured or operated so that residence time can be extended without extending machine length.

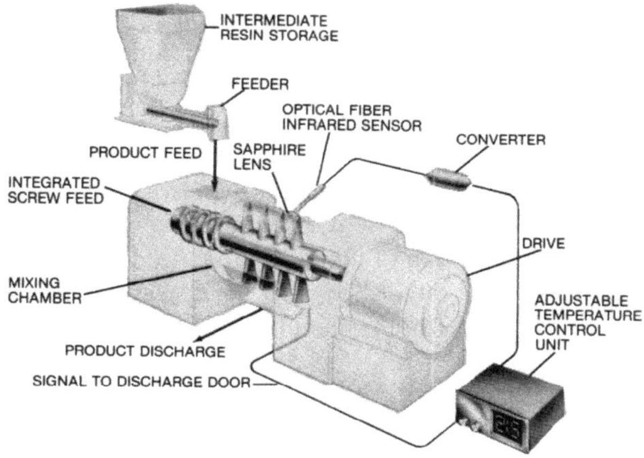

Figure 2.4 Modified intensity blender for batch-continuous plasticating of polymers, Gelimat Mixer (courtesy of Drais Co.).

In 1982 the Drais Co. (Mannheim, Germany) began production of a high speed cascading batch mixer called Gelimat under license from Carlew Chemicals (Montreal). The device consists of a horizontally mounted intensity blender with screw in-feed which dumps batches by alarm signal from an infrared thermometer (Figure 2.4). An intermittent batching feeder is slaved to the mixer and the discharged melt is transported to a die cutter by a discharge screw extruder or sheeted with a mill. The inventory in the discharge apparatus produces a continuous output. The mixer is primarily used to flux plasticized PVC.

In 1979, Tadmor *et al.* patented a disk processor on behalf of Farrel Machine and named the device Diskpack (Figure 2.5). The disks have superior drag characteristics to screws, are configured with blocks and pins, for spreading and mixing, and are manifolded together in series and parallel fashion to allow for more flexibility in surface exposure and local work input than on a screw device. Because the disk geometry has no helix, it causes no secondary flows. Recirculation flows and high stress clearance flows are predictably caused by placement of pins and dams. The Diskpack is used in various thermoplastic dispersion and devolatilization processes.

In 1987, Propex (Staufen, Germany) began producing the Conterna, a multiple chamber cascading device with two driven shafts per chamber (Figure 2.6). The chambers can be configured with combinations of kneading and transport elements. The shaft speeds in individual chambers are independently adjustable as are the restrictive dams between adjacent chambers. The Conterna was introduced to process highly filled materials where self-wiping and simple screw access are valuable; for example, in mixing energetic compounds.

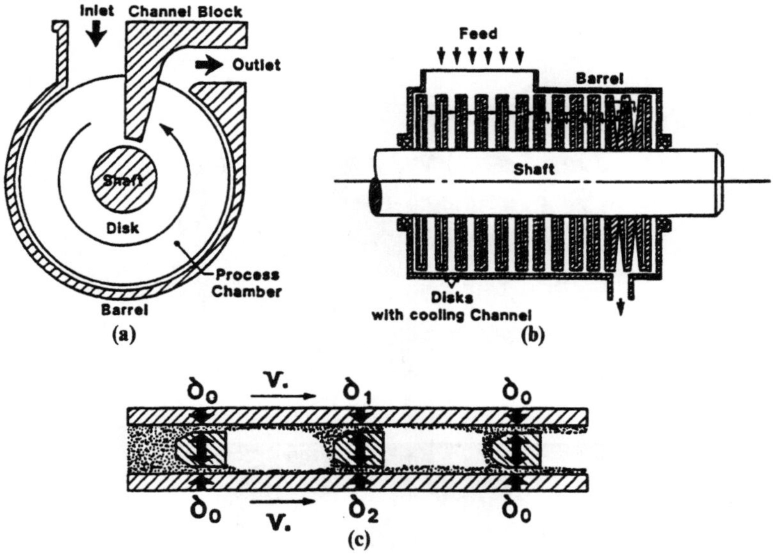

Figure 2.5 Typical configuration for Diskpack parallel disk plastic processor: (a) radial cross-section; (b) axial cross-section; (c) mixing pin arrangement.

Multishaft extruders are also available but are usually designed for specific processes. Four-screw devolatilizers and ten-screw reactors, for example, are known.

In the current continuous compounder market, more than ten manufacturers are supplying intermeshing co-rotating machines. JSW is competing with Leistritz in the supply of counter-rotating intermeshing compounders. JSW and Kobelco also produce continuous mixers with rotors.

The reader is encouraged to review White's excellent text [2] for a more complete history and mechanism of compounder design.

2.2 Mechanisms of dispersion

A logical progression in the technology of commercial blend preparation is from mechanisms of simple fluids deformed by simple machines to commercial fluids processed by commercial mixers. A simple fluid is purely viscous. Elastic effects, interfacial forces and phase change bring observable phenomena for which no fundamental theory exists. Convoluted mixer geometry, intersecting rotors, and large thermal gradients require sophisticated numerical analysis to reasonably predict conditions in the mixer. Prediction of a mixture's morphology requires a deformation–structure model for the fluid system and a compatible geometry–deformation model for the mixer.

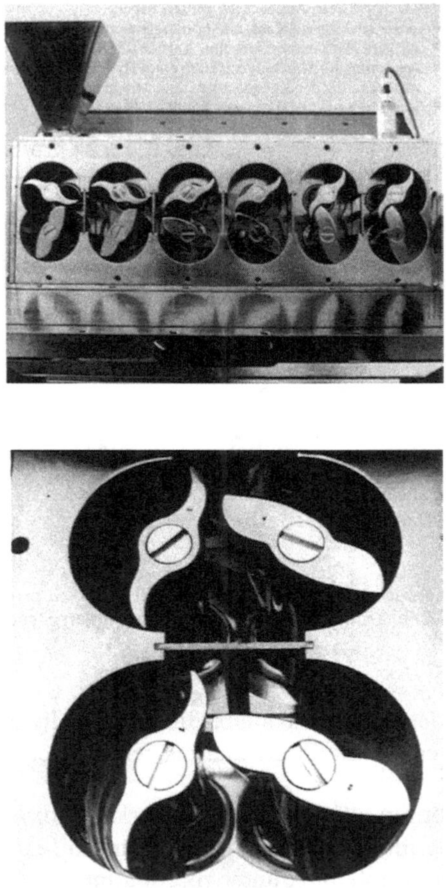

Figure 2.6 Conterna™ continuous mixer, a non screw-type compounder with stage specific residence time control (courtesy of Propex Co.).

2.2.1 Model studies

A natural place to start a discussion of blend technology is with liquid emulsions. In 1934, Taylor [3, 4] modelled the slow deformation and rupture of suspended drops in a dilute emulsion and conducted clever experiments to prove his theory. Key assumptions of his analysis are that:

- drop size is small compared to the apparatus
- the interface does not slip
- shear stresses are continuous across the interface (normal stresses are not due to the increased pressure within a phase because of surface tension).

By balancing viscous forces with surface tension changes defined by curvature of the suspended drop, Taylor developed an expression for the drop

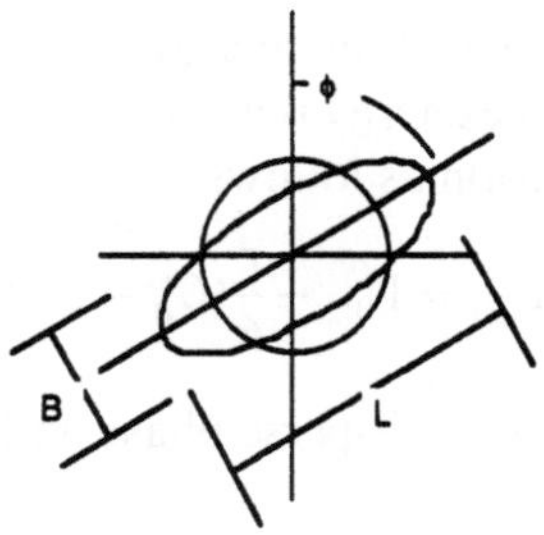

$$D=\frac{L-B}{L+B}=\frac{5(19p+16)}{4(p+1)\left[\left(\frac{20}{We}\right)^2+(19p)^2\right]^{1/2}}$$

Figure 2.7 Deformation of dilute droplets in a shear field after Taylor [4]. ρ = viscosity ratio, η_d/η_m where d indicates disperse phase and m indicates matrix. We = Weber number = $\eta\gamma r/\Gamma$, where η is matrix viscosity, γ is shear rate, r is disperse phase radius and Γ is the interfacial tension.

deformation (Figure 2.7). In addition, he defined a limiting droplet size for a given strain rate when viscosity and surface tension of the fluid phases are known. The break point occurs when the viscous forces acting on the drop surface overwhelm the stabilizing surface tension (Figure 2.8). Taylor's model was subsequently generalized.

Van Oene [5] argued that the recoverable free energy of deformation of a blend is proportional not only to disperse phase size but also to the difference

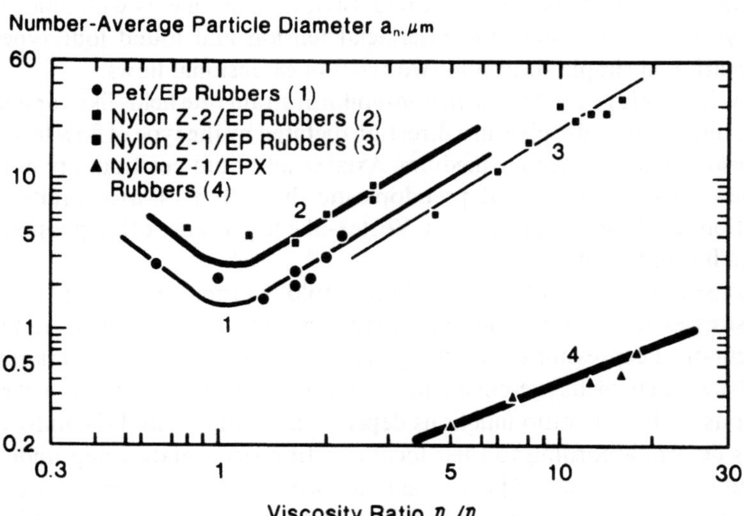

Figure 2.8 Effect of viscosity ratio on control of disperse phase diameter for blends with different interfacial strengths. The data can be combined into a master curve by plotting Weber number versus viscosity ratio (after S. Wu, *Polymer Eng. Sci*, **27**, 335, 1985).

in first normal stress difference of the phases, an elastic property of the fluids. Furthermore, the recoverable energy is a stabilizing quantity which should be added to the interfacial tension as follows:

$$\Gamma_{\text{eff}} = \Gamma_{12} + \frac{r}{6}(G'_{\text{d}} - G'_{\text{m}})$$

where G'_{d} = storage modulus of disperse phase and G'_{m} = storage modulus of matrix phase.

This has several familiar consequences:

(a) elastic droplets sustain higher critical strain rates before rupture than purely viscous drops.
(b) elastic fluids will constitute the discreet phase of the blend at higher volume fractions than viscous fluids.
(c) if the elastic phase is continuous, the viscous disperse phase will disperse easily.

Elmendorp and Maalke [6] studied blends of incompatible polymer melts and found that it is more appropriate to use a viscoelastic function in place of viscosity functions in Van Oene's development. Stable droplets survive 'super-critical' strain rates for polymer fluids. Time-dependent stability effects were found so that old drops were harder to perturb. Rapid shear rate changes disturbed this stability forming dispersions via Raleigh ripening (periodic disturbance and tip spawning of droplets). The phenomena were described in terms of the viscoelastic damping qualities of the melts.

Ramscheidt and Mason [7] repeated Taylor's experiments with fluids with a range of viscosity ratio and interfacial tension and found four types of deformation in simple shear and two types of extensional flows.

Flow in simple geometries with non-uniform strain rate fields like capillaries causes migration of droplets in a direction dictated by the type of flow field and the nature of the suspending medium. Axial equilibrium positions are normal for Newtonian medium and pseudoplastic drops but off-axis equilibrium conditions are found in other cases. Vortex-like flows cause other possibilities for equilibrium position.

A consistent theme in the study of morphology formation is the balance of fluid stresses to balancing interfacial tension and that concept degrades to mathematical expressions containing viscosity and first normal stress ratios. Non-Newtonian melts present an interesting anomaly to the definition of ratio functions in that the ratio functions depend on strain rate and the individual phases deform according to their local constitutional relationship under an applied load. Plochocki [8] neatly sidestepped the strain rate issue by developing the viscometric functions in terms of shear stress (Figure 2.9), a better conserved species.

Since a compounder's motion describes a strain field, an expression for a composite viscosity function is required to estimate an average fluid stress.

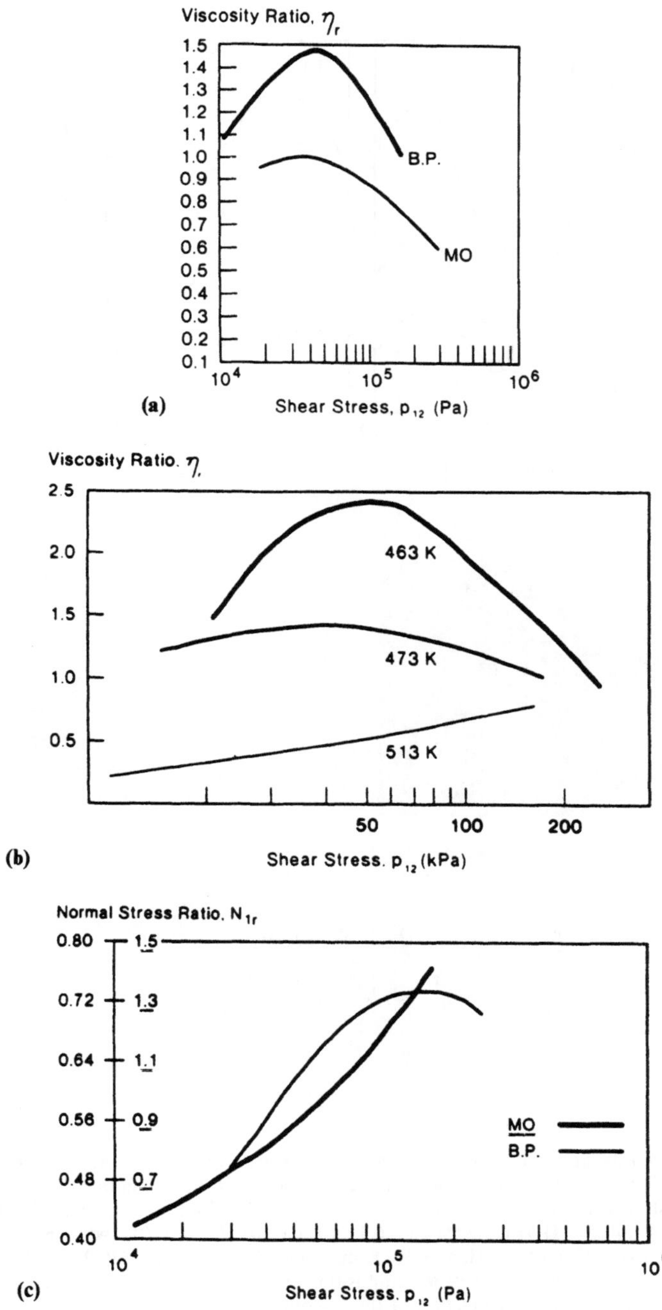

Figure 2.9 Rheological ratio functions for the polymer pair PS/LDPE, which are required for disperse phase size (and shape) estimation by a Weber analysis (after A.P. Plochocki *et al.* [23]. (a) Shear viscosity ratio function at 200°C for PS/LDPE supplied by British Petroleum (BP) or Mobil (MO); (b) sensitivity of viscosity ratio function to temperature; (c) first normal stress ratio function, an indication of relative elasticity.

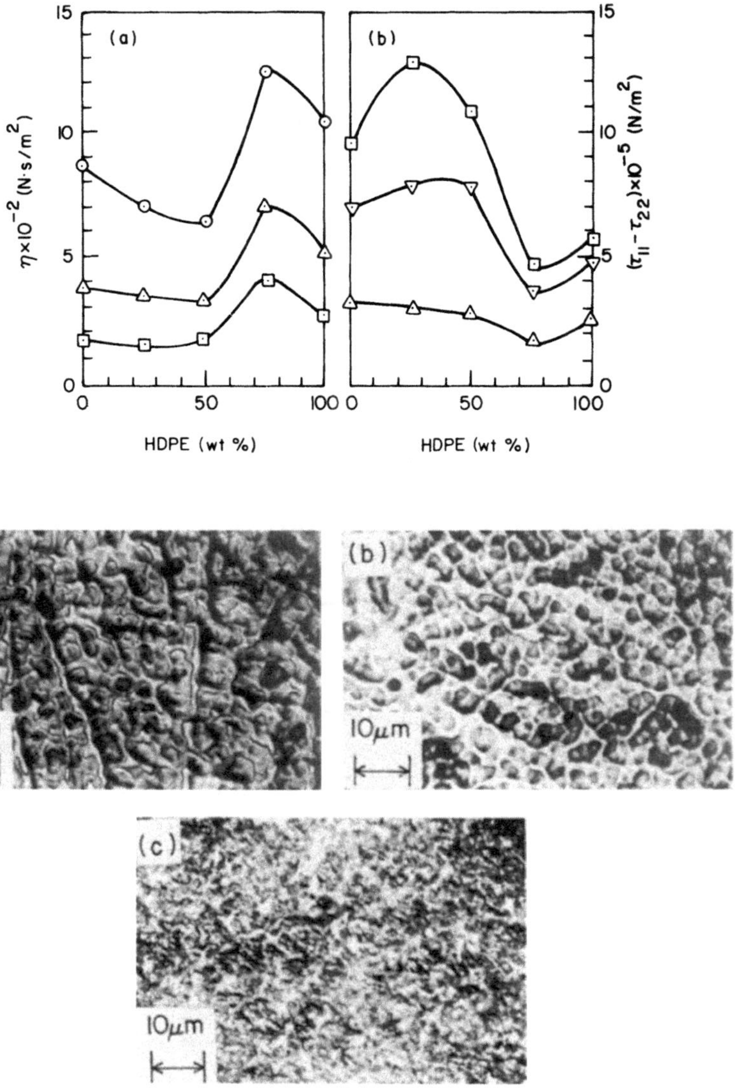

Figure 2.10 Viscosity and elasticity functions are not usually monotonic across a range of concentration. Shear viscosity and first normal stress function for the polymer blend HDPE/PS at 200 °C at wall shear stress of ⊙, 4×10^4; ∇, 9×10^4 and □, 10^5 Pa. Photomicrographs of extrudate cross section of (a) 25% HDPE (wt%) $\tau w=6.3\times 10^4$ Pa; (b) 50% HDPE (wt%) 6.1×10^4 Pa; (c) 75% HDPE (wt%) (from C.D. Han, Y.W. Kim, *Trans. Soc. Rheology*, **19**, 245, 1975). Note the antagonistic effect on shear viscosity at composition displaying a clearly disperse morphology and the synergistic effect at the co-continuous composition.

Rheological functions for polymer blends are seldom monotonically increasing across composition. More commonly, a minimum, maximum or even several extrema are found (Figure 2.10). Researchers have hybridized the concentration and shape effects found in suspension rheology with specific interfacial behavior and time-dependent phenomena like viscoelasticity and coalescence to explain this behavior. An excellent review is provided by Han [9] who also develops equilibrium flow profiles for stratified phases in terms of viscous energy minimization.

A significant factor in practical mixing is the effect of coalescence, a counter productive mechanism. Plochocki *et al.* [10] observed that dispersed phase size does not decrease indefinitely with work input from a mixer. In fact, minima of the function have been observed especially for blends with high interfacial tension, indicating a competing mechanism. The coalescence process requires collisions and is necessarily at least bimolecular while the dispersion theories are unimolecular (first order) meaning that coalescence dominates blends with a concentrated disperse phase. Serpe *et al.* [11] formalized the effect of concentration by proposing a modified capillary number as follows:

$$We^* = We(1 - 4 \times \phi_1 \times \phi_2)^{0.8}$$

where ϕ_1 = volume fraction of dispersed phase and ϕ_2 = volume fraction of matrix phase.

It would be derelict to assign the dispersion process exclusively to the melt zone of a plasticating compounder under mechanisms of fluid mixing. Several researchers [12, 13] conducted plasticating experiments with polymer blends in a small batch mixer and observed the condition of the mixture as plastication and mixing proceeded. They observe a multistep process which proceeded through suspension and/or semi-fluid and fractured material and/or dough-like composition before transitioning to a viscoelastic fluid (Figure 2.11). During plastication, dispersion quality is apparently affected by frictional changes brought about by embedding of particles of one phase in the still unmolten feed. Abrading, banding, fracture and kneading are some of the observed mechanisms which certainly control phase development.

Rotor configuration, feed sequencing, temperature policy, and operating conditions are factors that the technologist can adjust for optimal compounder performance. Fundamental dispersion studies can hardly predict the right geometry and conditions for a commercial process but they do identify useful rules, for example:

- Since dispersion is a phase reduction process, it pays to generate a well-mixed feed in the feeder train or feed section of the compounder.
- High stresses applied during melting are important and friction dependent, and care should be taken to reserve lubricating fluids, powders and low melting solids for post-addition.
- Compatibilizers reduce the interfacial tension and result in dramatically improved dispersions.

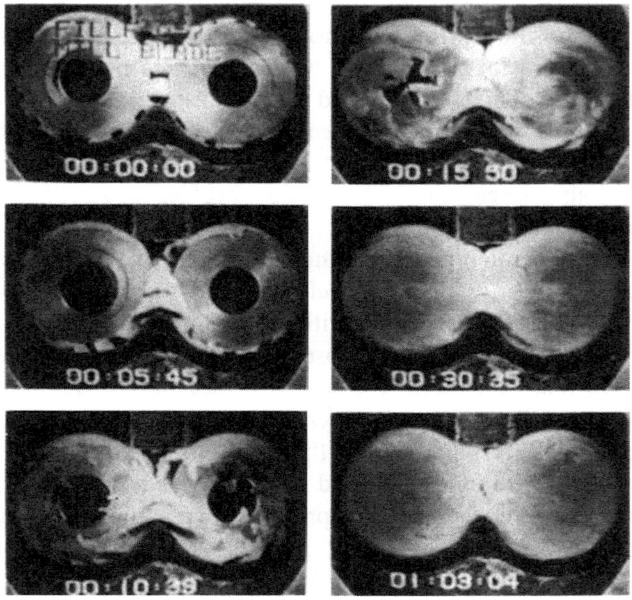

Figure 2.11 Plastification of EPDM in a laboratory Banbury type mixer progresses from tearing and banding of a doughlike substance to development of a viscoelastic fluid (from K. Min, *Intern. Polymer Processing*, **1**, 179, 1987).

- Stretching flows are more effective than shearing flows for phase rupture.
- Elastic melts resist deformation and may have to be compounded as concentrated suspensions at the co-continuous composition and subsequently diluted with matrix phase for adequate dispersion.
- Unless dispersions have been compatibilized with an interfacial agent they will tend to coalesce, especially for concentrated dispersions.
- Sustained deformation does not result in sustained phase dispersion. Rapid deformation changes cause more effective dispersion.

2.2.2 *Studies with mixers*

The phase mechanics of polymeric material combined with a tracking of position and orientation of all the droplets in a compounder would yield an exact simulation of the manufacturing process. The required computational resources would be immense. Mixing mechanics have traditionally been described for blends of immiscible fluids without interfacial tensions or, for that matter, any distinguishable physical features save an imaginary phase marker. In these analyses, velocity fields in the mixer are brought to bear on an initial orientation of phases bringing about deformation of phase boundaries. Droplet formation, *per se*, is not allowed. If the initial phases are stripes, the progress of mixing (increase of interfacial boundary surface) is measured by the change of frequency of phase boundaries through various cross-sections. Appropriate statistical measures relate to the overall spatial distribution and characteristic size of examined phases.

The influence of phase orientation on the direction of strain in the generation of interface is concisely discussed by Spencer and Wiley [14]. In their derivation, if disperse phases are randomly oriented, the interfacial area increases as a linear function of strain. If shear strain direction could be continuously oriented in a normal direction to the major axis of disperse phase, interfacial area would grow exponentially with strain. Extensional strains should be aligned with the major axis of the disperse phase. Because process chambers of mixers have walls, it is not possible to apply strains with arbitrarily large magnitude or orientation. Exponential mixing, therefore, represents a limiting case.

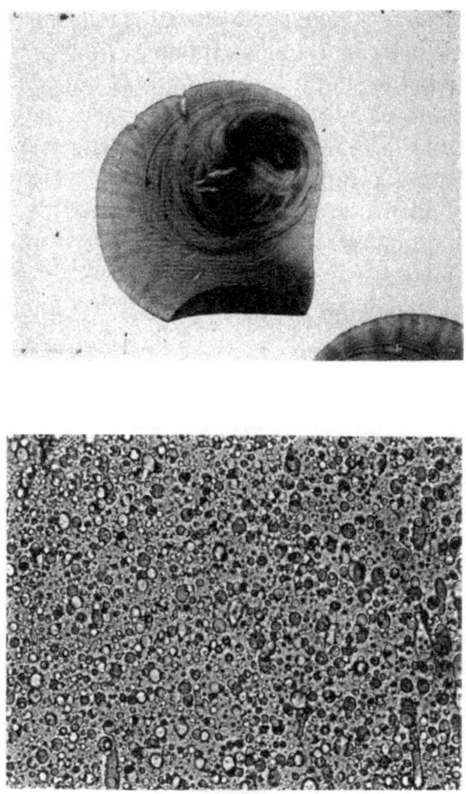

(a)

(b)

Figure 2.12 The morphology of the sample dictates the appropriate mixing analysis. (a) Texture analysis is described by propagation of interfacial area. (b) Disperse morphologies require particle tracking techniques coupled with chaos analysis.

If phases are described as particles instead of alternating bands (Figure 2.12), then particle tracking along streamlines describes development of texture. Reductions in computation time for this technique (and parametric representation of the result) is possible through the application of chaos concepts. Briefly, several sections are analyzed for evidence of large-scale interaction and activity in a section is analyzed for dispersing or agglomerating tendencies. From these analyses the efficiency of the mixer design can be estimated. The question of the minimum number of sections to be analyzed and the required similarity between sections is analyzed statistically. Chaotic tendencies do not eliminate possible formation of stable islands of agglomerates but experts in the technique claim a capability of forecast island formation by the character of the analysis.

Understanding laminar mixing of immiscible fluids with non-slipping interfaces is only a foundation. Researchers are studying the peculiarities of elasticity in melts and coupled time-dependent mechanisms like diffusion, coalescence, heat transfer and reaction for an enhancement of applied theory.

Gotsis *et al.* [15] have summarized several significant material–machine interaction phenomena by analyzing morphology from simple mixing devices. They performed studies in simple shear producing couette mixers, shear and elongation fields produced in eccentric couette mixers and abruptly changing fields in clearance areas with a drum and blade arrangement. Most strikingly, dispersions were more efficiently formed by the rapidly changing stress field of the clearance area than in a simple shear flow (Figure 2.13). These experiments qualify the spreading and scraping action of modern mixers for enhanced polymer–polymer mixing.

Consideration of gap flows has not escaped the interest of equipment

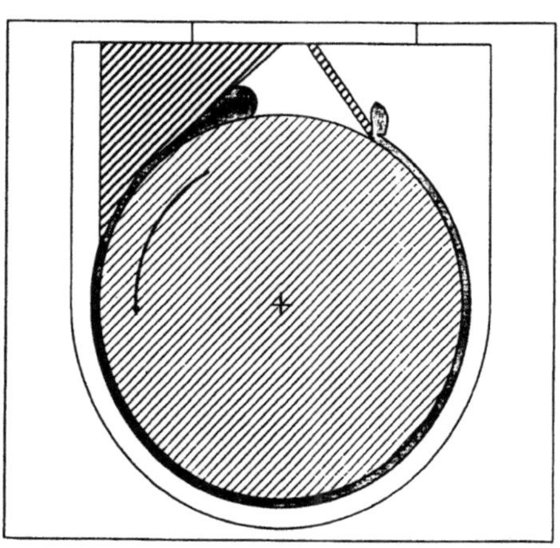

Figure 2.13 Gap flows are efficient in the development of disperse morphologies. Apparatus for generation of known number of gap passes (from A. Gotsis, B. David *et al.* [13]).

developers. Unfortunately, viscous flow through steep gap approaches has no analytical solution (like a lubrication approximation) and so researchers have resorted to numeric, statistical and experimental methods. Tadmor [16] introduced the concept of passage distribution function. Because of the absence of detailed flow analysis for mixers, the function relies on a simplified model whereby materials pass between stochastic (or well-mixed) and consecutive deterministic elements where exact deformation behavior can be modelled. Manas-Zloczower *et al.* [17] first used this method to describe carbon black dispersion in rubbers during Banbury processing. The probability of fluid particle passage through a high shear clearance area follows a Poisson distribution if the fluid is well-mixed from the time of the last passage. Manas-Zloczower attributed the dispersion process to the action of the clearance flow in which she developed orientation and stress dependent rupture dynamics.

Curry and Kiani [18] calibrated hollow glass spheres for their failure characteristic in fluid flow and used this behavior to determine experimentally the pass frequency for a Newtonian fluid through clearance areas of a ZSK continuous mixer. They found that clearance passage frequency depended on screw design, a significant (although expected) advance on the previous theory which related passage distribution to random events and fundamental equipment features such as the number of rotor tips in a mixer but not pitch or shape of the rotor. This study also showed a significant passage of fluid in clearance approach regions but not necessarily through the following clearance, indicating a geometric influence on the stability of rolling pools in gap approach areas (Figure 2.14).

David [19] investigated the effect of entry geometry to gap flows for various fluids. He found that elastic melts will form stable rolling pools in the gap entry unless the pool is perturbed by a periodic boundary or periodic flow rate.

The residence time experiment yields information about the overall backmixing capability of a mixer. In this technique, a detectable tracer is introduced to the feed as a pulse or as a step and the discharge stream is analyzed for tracer content over time. Data are represented in dimensionless terms (normalized by average residence time) and the variance of the normalized distribution indicates a tendency towards complete backmixing ($\sigma = 1$) or plug flow ($\sigma = 0$).

This simple analysis cannot describe spatial mixing but the results are important for processes with time dependency (like viscoelastic relaxation, diffusion, coalescence or reaction) and for the design of control systems. In order to develop physical insight into the process, investigators have compared their experiment against progressively complex models: CSTR, series and parallel tank networks, axially diffusing pipe flow, hybrid plug flow-diffusing models, etc.

Werner [20] compared the discharge tracer signal from an RTD experiment using a melt fed ZSK mixer containing kneading blocks to a network of tanks

(a)

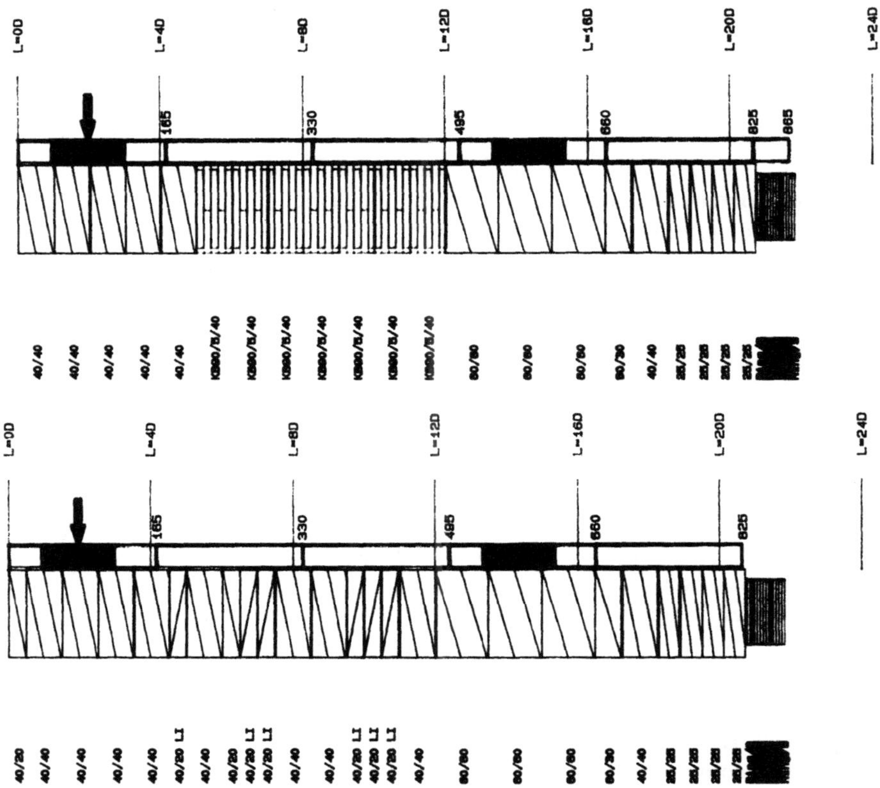
L=0D
L=4D
L=8D
L=12D
L=16D
L=20D
L=24D
165
330
495
660
825
865

(b)

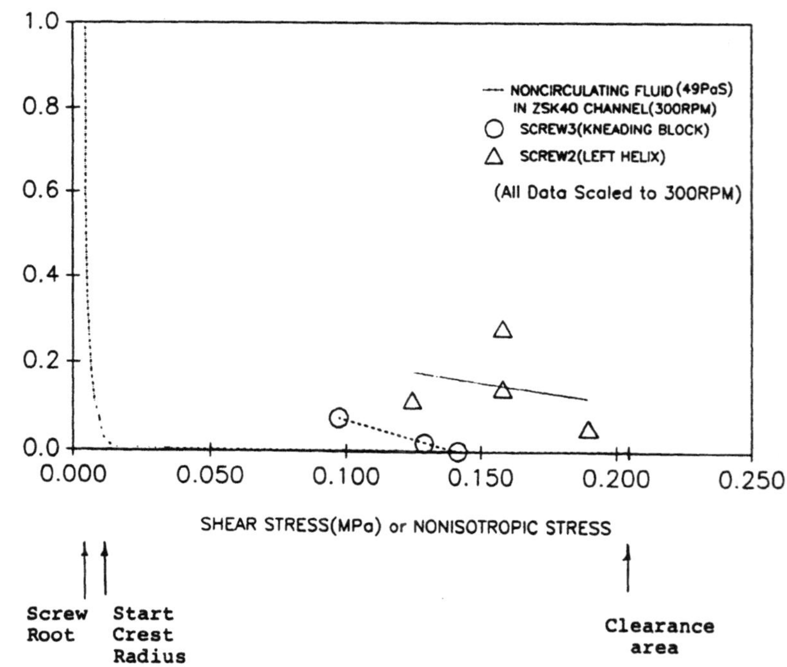
FRACTION OF TOTAL FLUID VOLUME
1.0
0.8
0.6
0.4
0.2
0.0
0.000
0.050
0.100
0.150
0.200
0.250
NONCIRCULATING FLUID (49PaS) IN ZSK40 CHANNEL(300RPM)
SCREW3(KNEADING BLOCK)
SCREW2(LEFT HELIX)
(All Data Scaled to 300RPM)
SHEAR STRESS(MPa) or NONISOTROPIC STRESS
Screw Root
Start Crest Radius
Clearance area

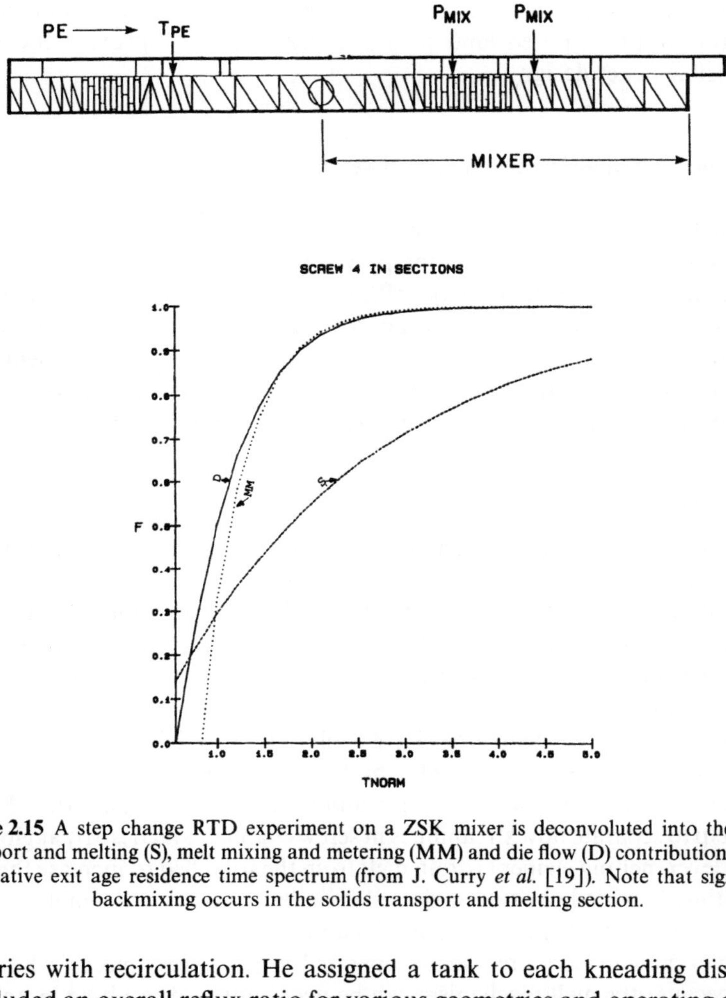

Figure 2.15 A step change RTD experiment on a ZSK mixer is deconvoluted into the solids transport and melting (S), melt mixing and metering (MM) and die flow (D) contributions to the cumulative exit age residence time spectrum (from J. Curry *et al.* [19]). Note that significant backmixing occurs in the solids transport and melting section.

in series with recirculation. He assigned a tank to each kneading disc and concluded an overall reflux ratio for various geometries and operating conditions by matching experiments to models. He found that reflux ratio increases with RPM and decreases with rate.

Curry *et al.* [21] investigated the damping capability of melt fed and plasticating ZSK compounders for the case of cyclic feed conditions sometimes found in plants. They found that backmixing was far more prevalent in the solid transport and plasticating zone than in the melt mixing zone (Figure 2.15). They also found improved damping performance in kneading blocks as compared to screw elements.

Figure 2.14 Experimental determination of gap passage frequency in ZSK mixer (from J. Curry and A. Kiani [17]). (a) Screw configuration, screw 3 is a kneading block intensive. Screw 2 is reverse helix element intensive. (b) Gap pass frequency is sensitive to screw design.

The combination of improving computational capability and skilled execution of the required fundamental experiments is improving the art of responsible mixer design.

2.3 Construction and operating principle of compounders

Although there are other techniques (solution blending, precipitation processes, dry mix and sinter) for polymer blending, melt mixing is most commonly used because of energy efficiency, operability, and environmental containment. The proven mechanisms for transporting viscous melts include drag and ram flow. In the previous section, the advantages of reorientation and gap flows for improved phase homogenization were detailed. These mechanisms should be embedded in a successful mixer design.

2.3.1 Single-screw

Although the single-screw extruder cannot generate the intersecting flows, pressure gradients or stress loading of a comparable twin-screw, the cost and installed capacity of these devices promotes their use as compounders. Very viscous amorphous melts and rubbers are commonly compounded on slow rotating pin barrel extruders. Preblends of fine powders with cohesive melt properties are also compounded.

Some unique screw geometries have evolved (Figure 2.16). The flight depth can be adjusted to accommodate the bulk of the feed stream, the solid to fluid compression ratio, and the pumping requirement of the die restriction. Barrel pins help to separate flow streamlines in screw channels and destroy agglomerates. Clever melting elements drain melt from the solid bed to increase the rate of melting. Dimpled cavities or slotted cylinders provide a damming restriction over which to disperse the contained melt mixtures.

Single-screw devices are usually operated at low screw speed (*c.* 120 rpm max) because the melt mechanism can be over-conveyed resulting in unplasticated product. Single-screws use simple thrust bearings which can be designed to operate at very high pressures (10 000 psi).

2.3.2 Kneaders

The Buss kneader is a reciprocating single-screw mixer which uses vanes and pins to produce a squeezing flow and vane wiping effect in the mixing zone. Vanes are interrupted screw flights. The barrel holds three rows of pins which wipe two sides of the vane during the pushing motion and the remaining sides during the retracting motion (Figure 2.17).

The vanes intersect the pins in a converging fashion so that extensional flows are generated. The clearances between pin and vane are close enough to

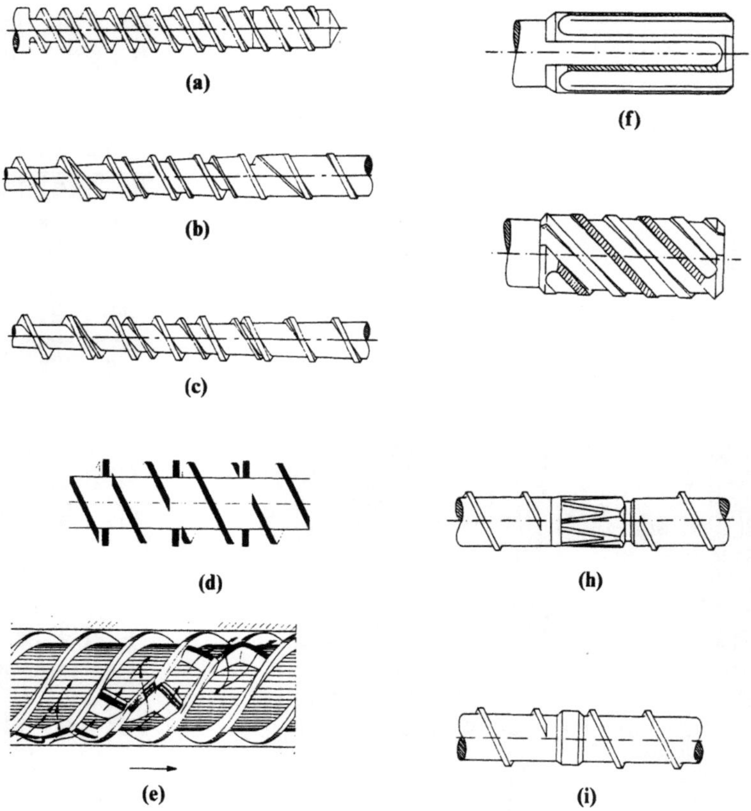

Figure 2.16 Adaptations of single screw extruders as compounders: (a) conventional single screw. Adaptations for improved melting performance: (b) Maillefer screw; (c) Barr screw. Adaptations for mixing of viscous melts: (d) pin barrel extruder; (e) EVK screw. Adaptations for improved melt mixing; (f) Union Carbide Mixer; (g) Egan Mixer; (h) Dray mixer; (i) Blister ring.

generate shear rates in the region of 2000 s^{-1} in the clearance gap at 300 rpm. The vanes are staggered so that the channel flow between a pair of vanes is split across the successive pair. The converging and diverging travel of vanes against pins causes a recirculating flow which is useful for phase reorientation. The kneader is usually followed by a discharge extruder and pelletizer.

2.3.3 Twin-screw

Relative rotation direction and extent of intermesh differentiate the various types of twin-screw compounders (Figure 2.18). Fully and non (tangential) intermeshing machines of the counter-rotating design are known. Only intermeshing co-rotating machines are built. An obvious advantage of a twin-screw

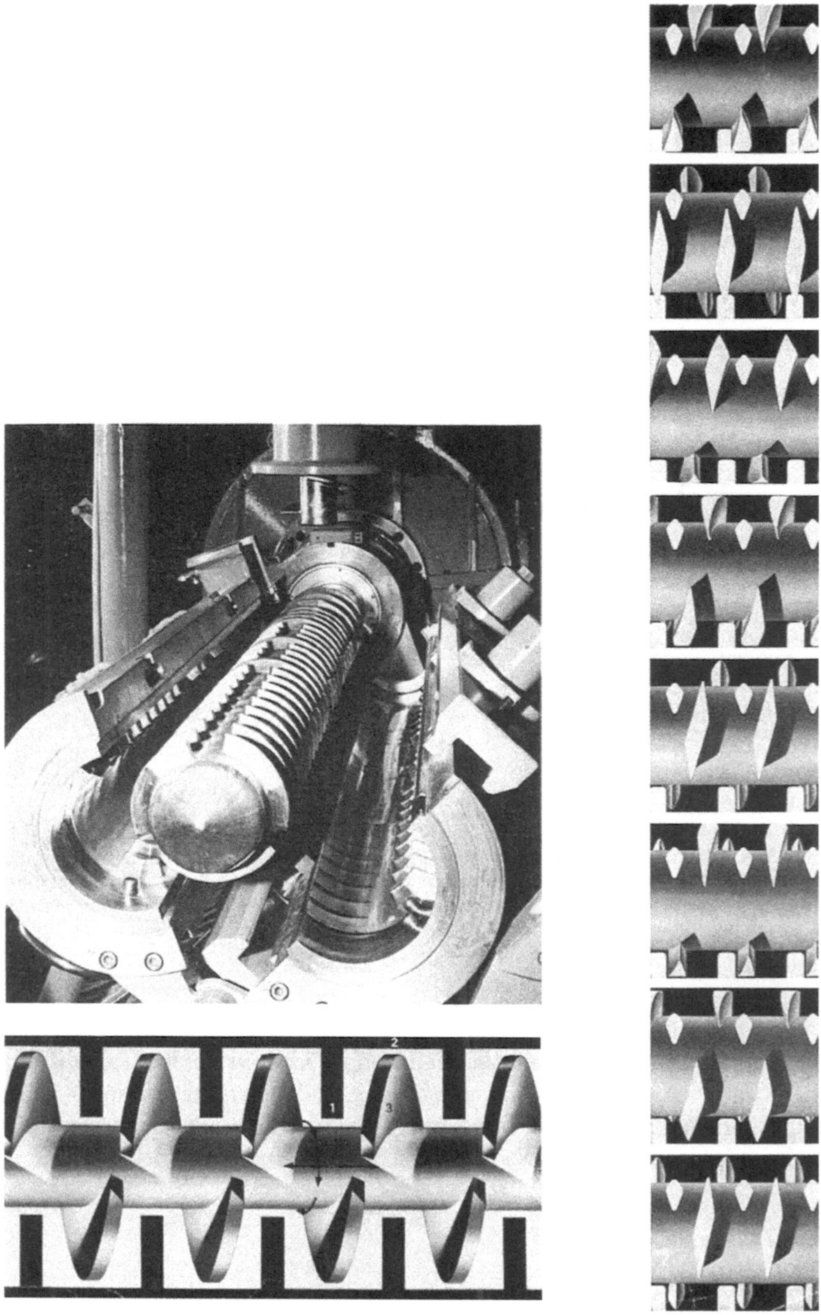

Figure 2.17 Buss Kneader reciprocating single-screw compounder. 1, kneading tooth; 2, kneader casing; 3, screw flight.

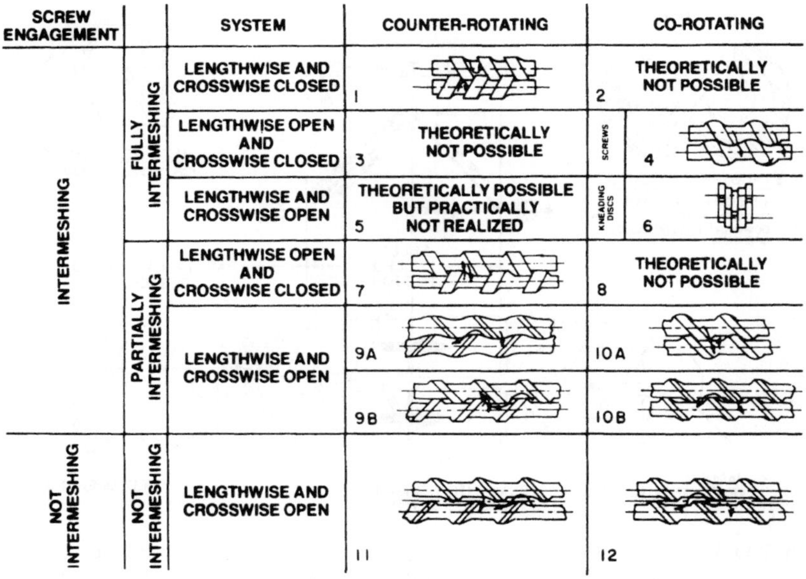

Figure 2.18 Classification system for twin-screw extruders (from K. Eise *et al.*, *Advances in Polymer Technology*, **1**, 1, 1981).

design over a single is the additional distance along the periphery over which the melt is dragged in one revolution. Less obvious features distinguish these machines as polymer compounders. The cross flow between screws causes efficient phase orientation and stress peaks. Intermeshing machines are typically designed as a self-wiping geometry. Counter rotating machines can be positive displacement pumps. Twin-screws are starve fed so that components of a formulation can be fed instead of preblending them. Machines are easily configured for multiple-feed addition and vent stages.

2.3.4 Counter-rotating tangential

These machines, built exclusively by Welding Engineers, consist of two parallel single screws, tangentially mounted and rotating together at the top (Figure 2.19). Since these machines are not constrained by the tolerances required of intermeshing machines, they can be built to practically arbitrary length and 120/1 (L/D) machines have been built. (Intermeshing machines are limited to *c.* 50/1 by machinery tolerances.) The screw flights can be matched at the apex for a better pumping behavior or offset for better backmixing behavior. Blister rings, which choke the extruder flow and cause high shear stresses when overcome, can be mounted side by side to completely block the flow channel. One shaft extends at discharge to meter melt to the die as a conventional single-screw.

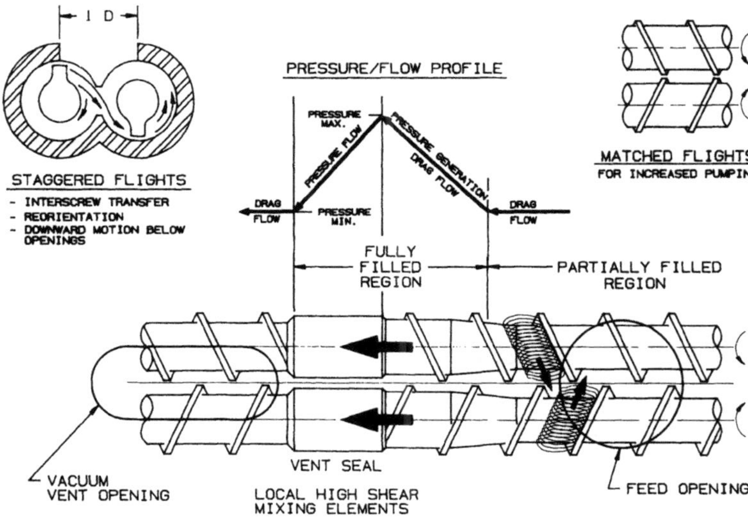

Figure 2.19 Welding engineers non-intermeshing counter-rotating twin-screw extruder (courtesy of Welding Engineers Inc.). Note: Matched or offset configuration, variable flight depth, blister rings.

The success of the tangential twin-screw derives from the dynamics of the interscrew apex area. For feeding, this apex increases the volume into which feedstocks can be dragged into the extruder. For venting, it is convenient to drag bulging melt away from the vent opening, and crush foams to release volatiles without a restrictive intermesh. In gas-melt processing, the apex is a plug area through which flows are convected without the destabilizing effect of a passing flight. Staggered flights provide an efficient mechanism for stream splitting. The screw flights generate a pressure by which some fluid is pushed across the apex. The amount of fluid depends on whether it sees open space (staggered flights low fill) or a full and possibly pressurized flight (matched flights or any flight with high degree of fill). Because the transferred flows are directed against one another at an angle, phase orientation results.

2.3.5 Counter-rotating mixers

The Farrel Continuous Mixer is a hybrid mixer compounder which, broadly speaking, is a counter-rotating non-intermeshing compounder. A screw feed supplies the plasticating and mixing rotors. The effect of the rotors can be changed by regulating the FCM discharge value which has an effect somewhat analogous to the ram pressure of a Banbury. Compounded stock is discharged to a screw or gear pump and pelletized.

2.3.6 *Counter-rotating intermeshing*

These devices are derivatives of positive conveying screw pumps. As such, they have advantages for die pressurization and for pumping low viscosity fluids. Without modification, however, the original design would continually recirculate the same fluid volume except for leakage flows over crests or through the tetrahedral space at flight intersection, and would discharge melt in pulses. These deficiencies in mixing behavior and pumping have been relieved by design modifications.

The channel shape for intermeshing counter rotating machines is defined by rolling discs against one another. Such a translation generates a cycloidal gear shaped channel and wide crests (compared to co-rotating machines) (Figure 2.20). The flight is often cut as a tetrahedron compromising some of the positive displacement to leak flow and reducing the crest width allowing a gap for cross-channel flow and increasing the process volume. The milling action of the intermesh can be used to advantage by employing undersized elements or special interleaved blister rings (Figure 2.21).

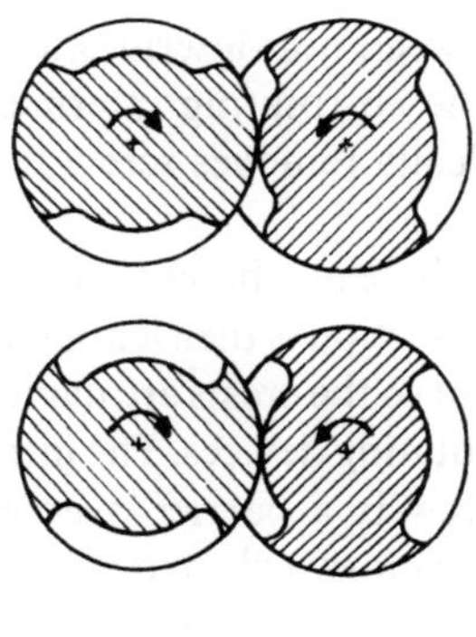
(a)

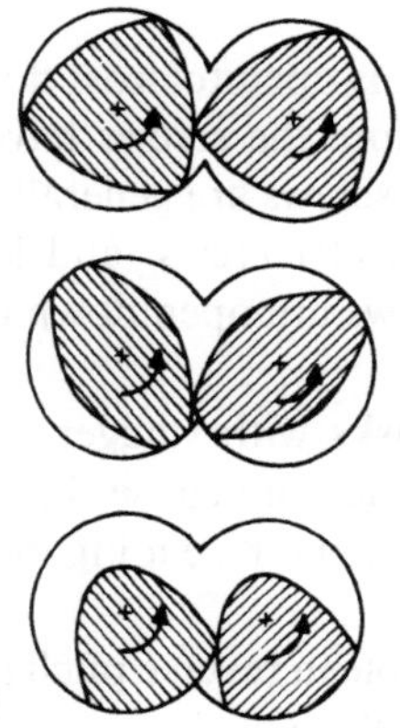
(b)

Figure 2.20 Cross-sectional screw shapes for fully intermeshing extruders: (a) cycloidal shape of counter-rotating mechanism; (b) Lenticular shapes of co-rotating machines.

Figure 2.21 Intermeshing counter-rotating extruder (courtesy of American Leistritz Extruder Corporation). Note: cycloidal channel shape, variable pitch and overlapping blister rings.

2.3.7 Co-rotating intermeshing

A hydrodynamic hybrid between non-intermeshing devices and the positive conveying screw pump is the co-rotating intermeshing and self-wiping extruder. Self-wiping mechanically removes screw residue which would otherwise retard fluid flow.

The channel geometry for these machines is defined by sliding disks around one another. The screw cross-section is then lenticular in shape. At the crest the flight walls are still convergent so the crests are narrower and the free volume is larger than for the counter-rotating devices. The intrusion of one flight into the channel of the opposite screw is a barrier over which material must flow causing a twist restraint which forwards material directly down the machine when overcome. The same restraint causes a restriction to pressure flow from the die or high pressure areas.

The co-rotating geometry is open to cross flow at the intermesh except for the wiping flight and therefore the screws have an open path from feed to discharge (Figure 2.22). The screws are generally cut with 2 or 3 starts. Fewer flights (or starts) result in wider crests (and lower free volume) at a fixed diameter. Although the screws are open down channel, inter-channel flow is restricted by the flights.

Collections of profiled disks with stagger angle, called kneading blocks, open the cross-channel contact and cause the stresses with their wide crests that plasticate and mix polymers. Kneading blocks maintain the self-wiping profile.

Screw elements are available in reverse pitch for maintaining back pressure dynamically against mixing elements.

Successful mechanisms found in other compounders have been emulated on the co-rotating machines. Large clearance rotors have been applied to olefin

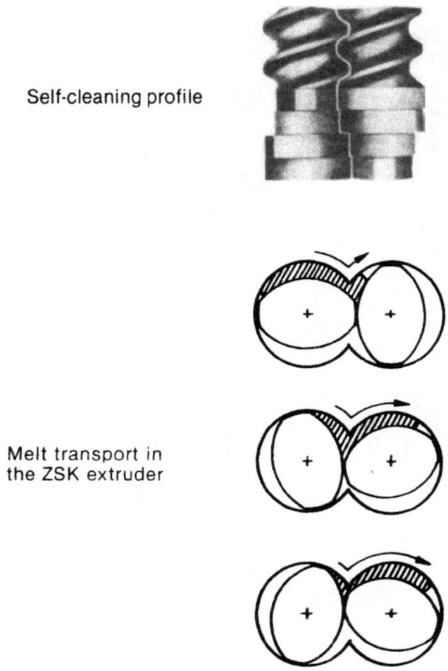

Figure 2.22 Fully intermeshing profile and material transport in a co-rotating twin-screw extruder for screws and kneading disks.

and rubber processing. Interleaved disk compounding elements and barrel valves have been used to improve difficult dispersions.

The transport efficiency of the self-wiping geometry is the nemesis of certain processes especially diffusion limited types like solution forming, thermal equilibration or plasticizing. A series of non-self-wiping elements has been developed for these processes. Slotted screw elements, turbine shaped elements and turbine shaped elements cut on a helix, have proved successful (Figure 2.23). In the compromise between adequate machine volume and adequate strain history, novel mixing elements have been designed. Eccentric lobed discs and multilobed discs are effective tools for reliable melting in high volume machines.

2.4 Comparative testing

The mixer–material interface is a hot bed of technical investigation. Rapid progress is being made in simulation, experimentation and interpretation of results. Consortiums have been formed to investigate processing behavior and product qualities of materials and mechanisms.

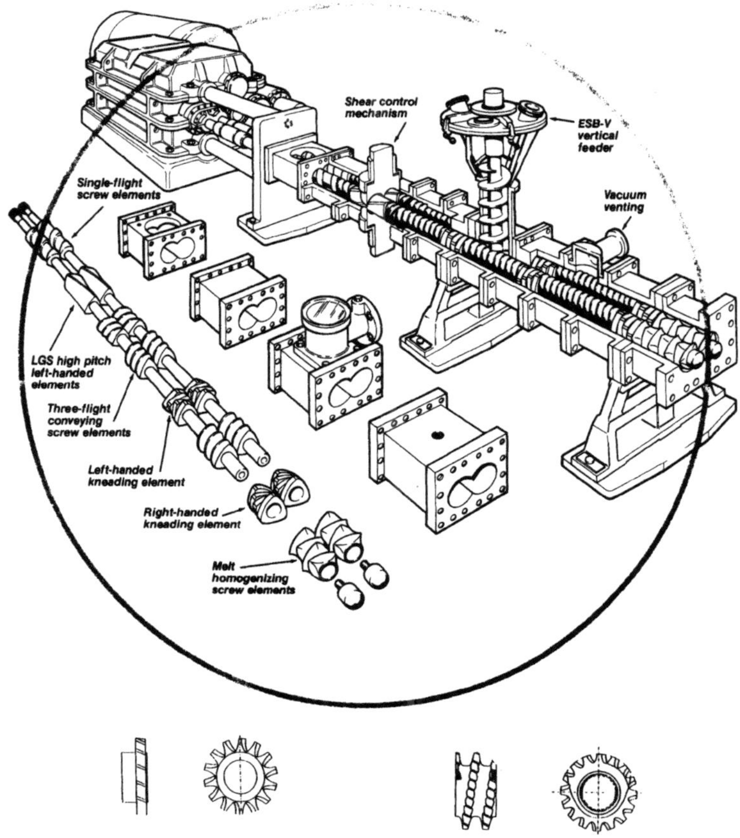

Figure 2.23 Co-rotating intermeshing extruder type ZSK (courtesy of Werner & Pfleiderer Corporation). Note: Modular design. Disk-type kneading blocks and melt homogenizing elements of a non-fully wiped design.

Polymer blends can be broadly classified as incompatible, compatibilized, or miscible. Plochocki [22] organized a study of phase morphology of incompatible and compatibilized blends processed on each of five prevalent compounders. The participants subsequently published their results [22–26]. The machines tested were Werner & Pfleiderer's ZSK-30, Berstorff's ZE-40, Baker Perkins' MPC/V50, Farrel's Diskpack and FCM4, and Buss Condux's Kneader. The ZSK-30, ZE-40 and MPC/V50 are co-rotating intermeshing and self-wiping design. The most distinguishing geometric feature between these machines is a progression in relative flight depth (depth/diameter) and decrease in shaft spacing from ZE-40 to ZSK-30 to MPC/V50, respectively. Each manufacturer favors a mixing element design in the study, for example Berstorff tests gear mixing elements, WP the traditional kneading blocks, and BP the highly staggered kneading discs. All manufacturers feature modular construction of barrels and screws and therefore the tested designs are somewhat arbitrary. Considering the modularity of machine design, there is potentially a lot more overlap in capabilities than the following data represent.

The chosen process materials were low density polyethylene (LDPE) and polystyrene (PS). This pair have several desirable properties. The proximity of major phase transition minimizes crystallization and solid state strain effects. PS is easily solvent extracted for high contrast phase differentiation. The components have a viscosity ratio close to 1 in an engineering range of shear stress so fine dispersions should be feasible (Figure 2.9). PS, the disperse phase, is more elastic than LDPE at high stress, so simple spherical morphologies should be attainable if high stress can be developed. Flow activation energy differences are substantial so PS viscosities are degraded much more substantially than LDPE thereby allowing a viscosity ratio depending on melt temperature. On the other hand, the interfacial tension between PS and LDPE (around 5 dyne cm^{-1}) is enough to encourage appreciable coalescence. Alloys were also prepared through addition of 5 pph S-EP copolymer compatibilizer, thereby reducing interfacial tension.

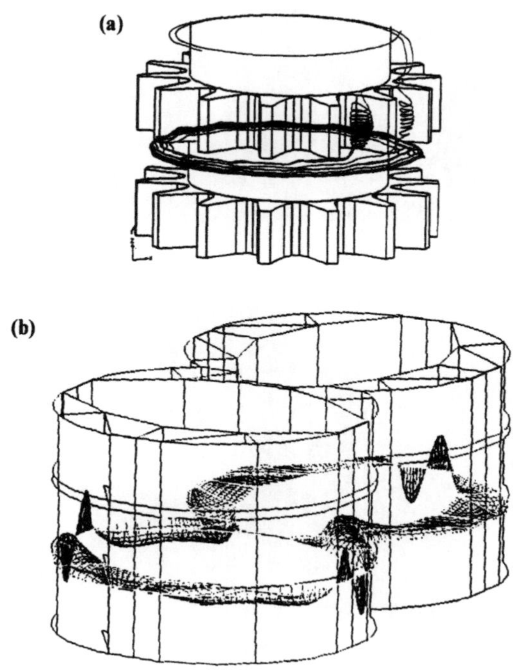

Figure 2.24 Computational flow field representation for two geometries in ZSK compounders (from A. Kiani *et al. Proc. 50th SPE Antec* (in print), 1992). (a) Particle tracking in turbine mixing element; (b) axial velocity field at intersection of successive disks in a neutral kneading block arrangement.

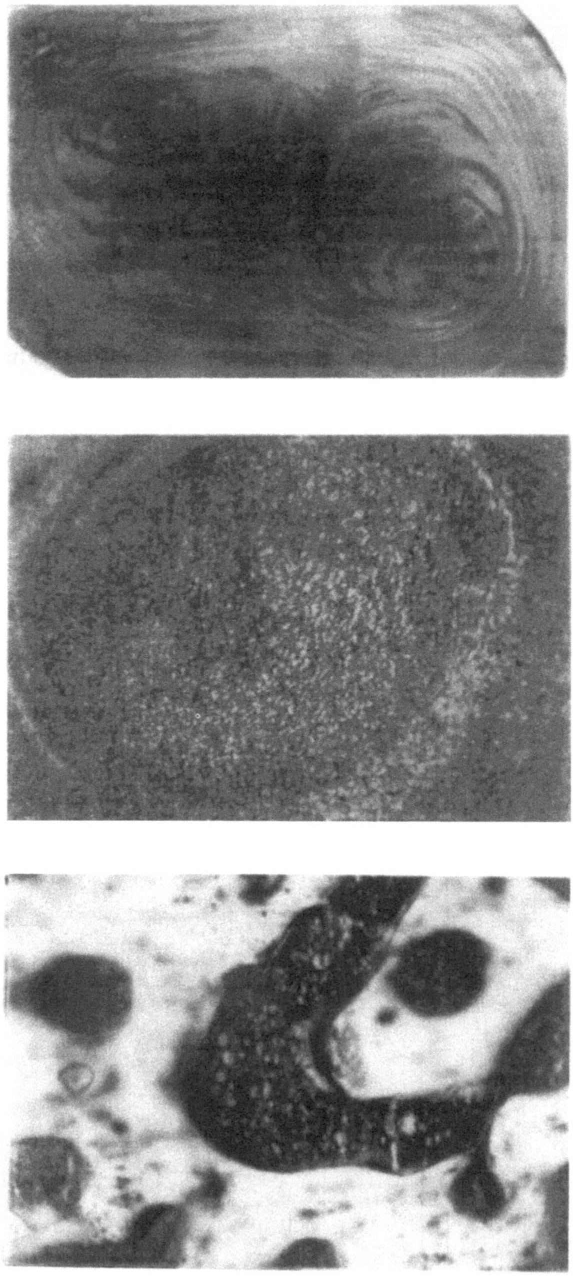

Figure 2.25 Typical large-scale and time-dependent morphological features for PS/LDPE system (from A. Plochocki *et al.* [23]).

Inasmuch as each compounder has distinguishing mechanical and geometric features, the morphology and interface development should be responsive to the particular distribution of process history, process time and plastication mechanism (Figure 2.24). Off-setting these differentiating effects are die flows, which tend to phase segregate material, and differential quench rates which promote viscous phase relaxation and coalescence (Figure 2.25).

The standard composition contained 33% PS. In all cases PS formed a disperse phase usually shaped as an ellipsoid aligned along the direction of flow. Alloys always formed finer and less disperse morphologies than the blends. After tedious examination of microphotographs from specimens, cut in both radial and axial cross-section the average morphology could be described as an ellipsoid with 4:1 axis ratio (Figure 2.26). Phase size was gauged as the diameter of the radial section. All mixers had an optimal specific

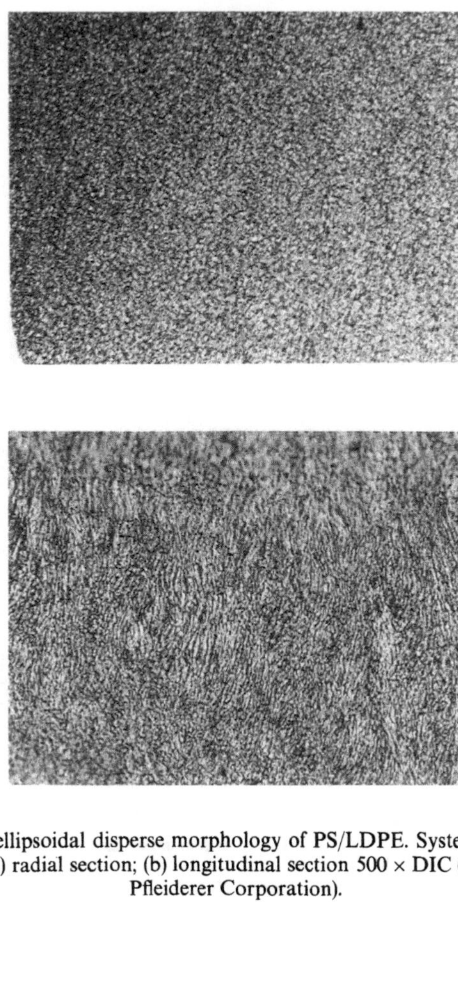

(a)

(b)

Figure 2.26 Typical ellipsoidal disperse morphology of PS/LDPE. System of quenched strand from compounder: (a) radial section; (b) longitudinal section 500 × DIC (courtesy of Werner & Pfleiderer Corporation).

mechanical energy input at which to produce the finest morphological features (Figure 2.27).

2.4.1 The WP study

WP produced blends both in single train plasticating mode and in a cross-head arrangement used for blending melts of the components using the laboratory scale ZSK-30.

Conceptually, melt blending should produce the most unambiguous result. Melt mixing energy and melt mixing time were experimentally decoupled from the plasticating process in the experiment. Plochocki *et al.* [10] considered the formation of a disperse phase under these conditions. Considering a 2 μm diameter dispersion of 4:1 ellipsoids (a fine dispersion for a blend), the volume fraction of the PS phase, and an experimentally measured interfacial tension of 5.4 dyne cm^{-1}; only 3.24 kJ m^{-3} are required to generate the required surface. The other 0.15 MJ m^{-3} are spent in transporting and heating the

(a)

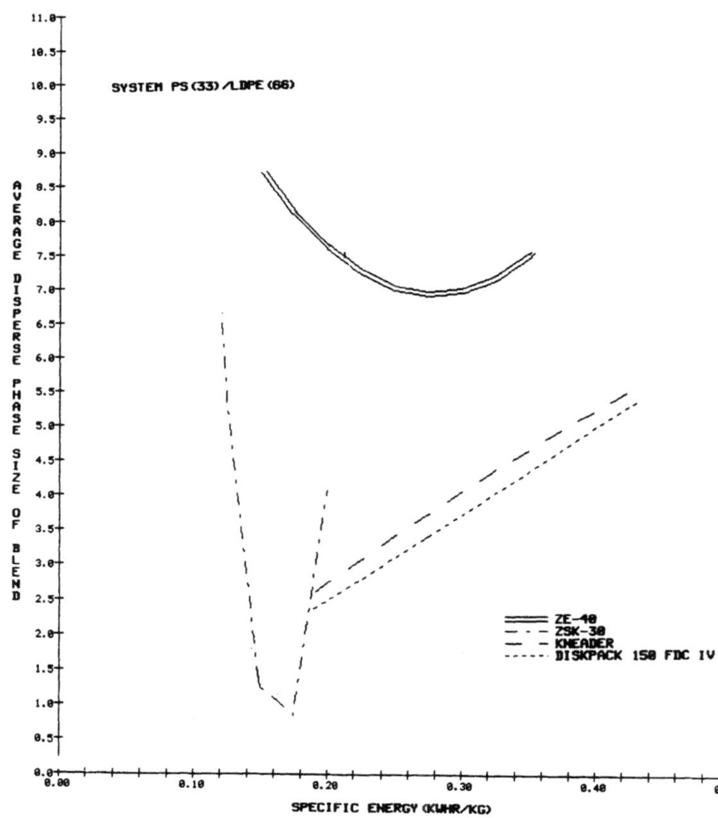

Figure 2.27 (a)

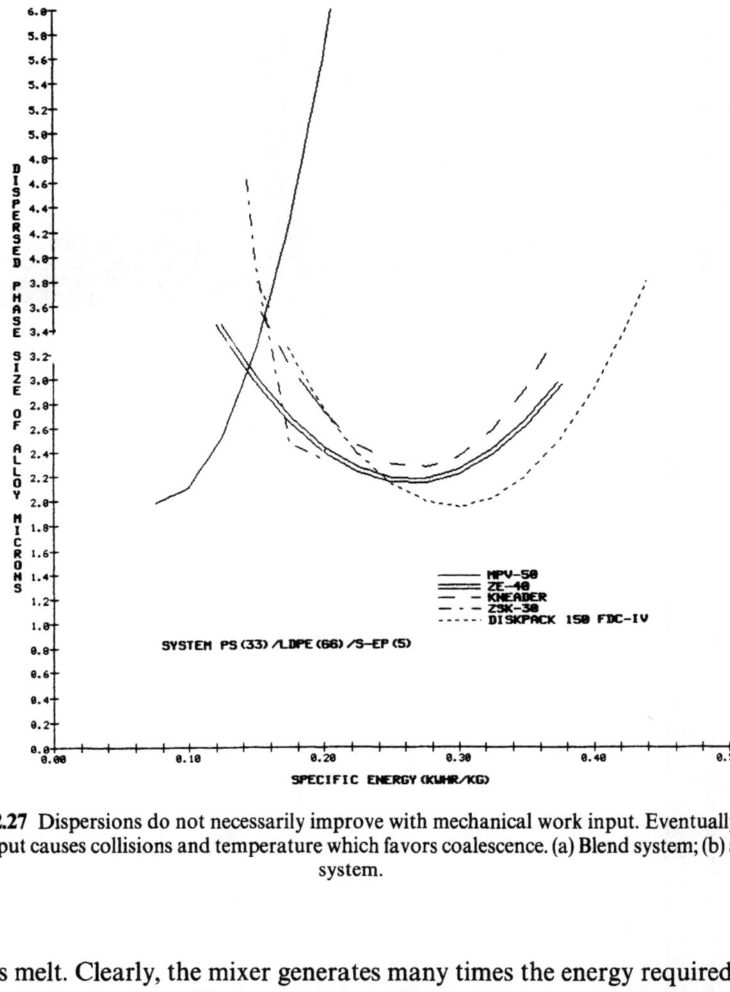

(b)

Figure 2.27 Dispersions do not necessarily improve with mechanical work input. Eventually the work input causes collisions and temperature which favors coalescence. (a) Blend system; (b) alloy system.

viscous melt. Clearly, the mixer generates many times the energy required for surface generation but effective stress transfer between phases, viscous damping of phase deformation and coalescence limits the dispersion process. Plochocki *et al.* further argue that at the compatibilizer concentration of 5 pph, and again assuming a 4:1 ellipsoid of 2 μm diameter, the compatibilizer exists as a physical layer of *c.* 6 monolayers. This interphase could dissipate by diffusion into the major phases leaving a monolayer and optimally compatibilizing the alloy. Regardless of interfacial layer thickness, the interfacial tension is reduced by the addition of compatibilizer and was measured at 0.94 dyne cm^{-1}, resulting in finer and lower variance distributions of disperse size for compatibilized alloys (Figure 2.28). Finally, coalescence is known to be an operating mechanism since phase size at similar process conditions is concentration dependent.

Rate and screw design have noticeable effects on morphology. The rate has a

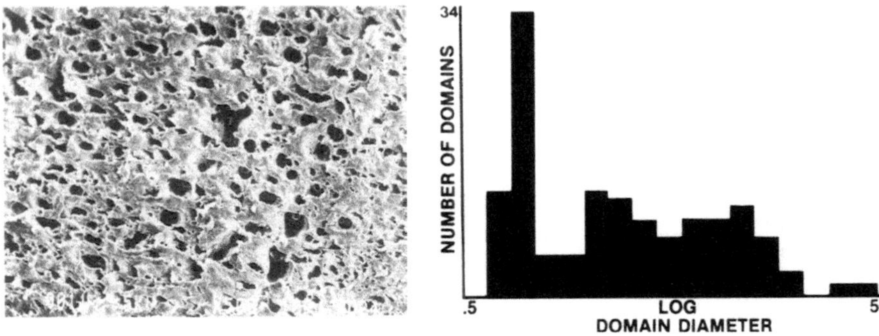

(a)

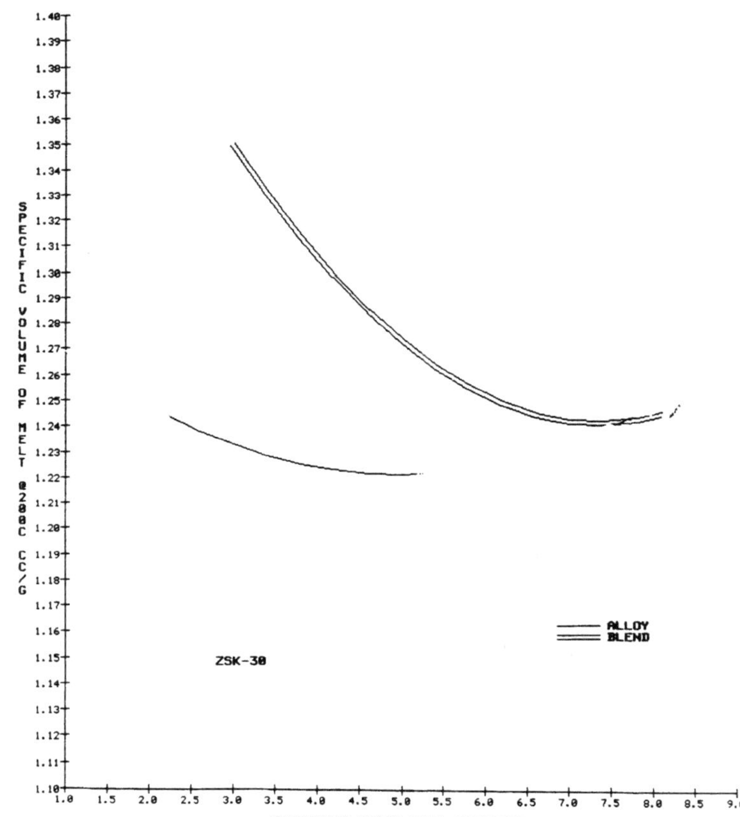

(c)

Figure 2.28 Reducing interfacial tension through the use of copolymers. Compatibilizer improves the dispersion process (from A. Plochocki *et al.* [23]). (a) Blend of PS/LDPE (33%wt PS); (b) same composition with 5% of S-EP copolymer. (c) The quality of the interface affects melt volume. Dispersion process in blends with high interfacial tension can lead to interfacial voiding. 500 × DIC (courtesy of Werner & Pfleiderer Corporation).

(b)
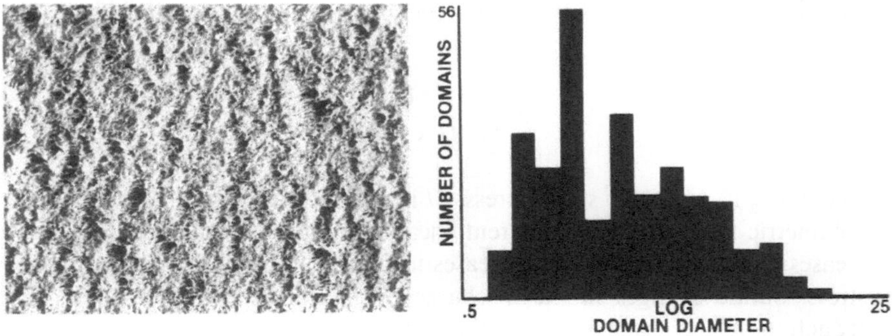

Figure 2.28(b)

(a)
(b)

Figure 2.29 Plasticating effect on the dispersion process for PS/LDPE systems: (a) sample produced by melt mixing; (b) sample produced by single-train plasticating compounder.

straightforward effect on regulating the stress loading onto the blend as expressed by:

$$P_{12} = \frac{U}{Q \times V \times t}$$

where P_{12} is apparent shear stress, U is specific mechanical energy, Q is volumetric flow rate, V is apparent shear rate and t is residence time. Decreases in rate correspond to increases in specific energy and average shear stress. Subtle changes in the discharge configuration are responsible for reduction of large-scale texture in the melt.

In comparison to the melt mixed specimens, the materials prepared on a plasticating compounder had finer and more complex morphologies often containing phase within phase structures (Figure 2.29). Certainly, the plasticating process contributed to the dispersion with possibilities for abrading solids, semi-fluid fracture and tearing mechanisms, and localized phase inversions.

Specific melt volume is an indicator of the quality of the interface. The specific volume increases with finer dispersions as the imperfect interfacial area grows. For the blend, specific volume grows faster than for the alloy. This behavior reflects the low interfacial bond strength of the blend which is subject to failure during processing.

2.4.2 Berstorff

The ZE40 pilot plant compounder was arranged in three configurations [24]. A standard plasticating zone was followed by either a kneading block melt mixing zone or a gear like mixer. The kneading block screw was modified as a variant which accommodated downstream feeding of the minor component, PS, into the LDPE melt.

Gear mixers generate surface by stream splitting which does not generate the mechanical energy or melt temperature of the shear intensive kneading blocks. The morphology responds more effectively to kneading blocks than gear mixers indicating the need for the high shear and stretch rates available in kneading blocks (Figure 2.30).

The downstream PS addition deprived the process of the residence time and dispersive stresses available in the melting process of the plasticating compounder. The morphology is correspondingly coarse.

Melt temperature control for viscous melts is predominantly a function of RPM. Two effects of increased temperature are change in viscometric ratio functions and acceleration of coalescence. For the model system, the shear viscosity ratio changes from 1.4 to 0.45 and the first normal stress ratio falls in a like manner at higher melt temperature. The higher temperature process results in appreciable coarsening of the dispersion.

(a) (b) (c)

Figure 2.30 Configuration and feed technique effects on morphology for PS/LDPE system. (a) Turbine type mixing element used for melt mixing; (b) kneading block mixing elements used for melt mixing; (c) kneading block mixer with disperse phase post added into LDPE melt (after M.E. Mack *et al.* [22]).

2.4.3 Baker Perkins

Baker Perkins uniquely reported only on the compounding of compatibilized alloy in their MPV-50 [26]. These data show the unique tendency of phase coarsening with increased specific energy. The larger phases also seem more spherical in shape. The data also show a reduction in standard flow rate (increase in viscosity) and decrease in melt swell with increased specific energy. The other manufacturers report a reduction in phase size with specific energy or, in cases where very high melt temperature accompany high specific energy, an optimal value for fine dispersions because of reduced coalescence tendency compared to the blend.

Remember that the MPV is a deep flighted machine (as compared to ZSK or ZE) which is reflected in the low specific energy range reported. The machine uses a unique barrel valve to force melts through clearance areas as well as high rpm to generate energy. It is possible that the morphology generating stress happens early in the alloying process and that a relaxation process proceeds in the deep flighted discharge element wherein the effect of additional specific energy is to provide more residence time, increase phase collision, and improve the distribution of the interfacial layer. Decreases in blend viscosity and die swell behavior are consistent with improved interfacial adhesion.

2.4.4 Farrel

Results were reported for Diskpack Model 150 FDC-IV [27], a compounder consisting of six parallel melting chambers, arranged in series and with pins, a low pressure mixing chamber and a high pressure extrusion chamber (Figure 2.5). Dispersive phase coarsening was found for both blends and alloys with increased rotor speed, melt temperature or specific mechanical energy.

The Diskpack is a unique device because of the improved drag flow efficiency of a disk chamber and because process chambers are connected with manifolds. Manifolding the chambers permits the designer to choke specific process chambers, for example melt mixing chambers, and decouple the rpm–throughput relationship of conventional screw-type compounders. Higher shaft speeds may change viscous dissipation and melt recirculation without significantly affecting residence time.

It is known that the model system will coalesce in quiescent flow situations, and at an accelerated rate under mild agitation. The Farrel results again show that morphology development owes a lot to the plastication process and that indiscriminate melt mixing may actually coarsen the structure.

2.4.5 Conclusions to the mixing study

A knowledge of some fundamental mechanisms which drive the dispersion process can be beneficially applied to optimizing commercial compounding processes. Choice of compounder, operating condition (rate, rpm, temperature), feeding technique and compounder configuration have effects on the quality and texture of the resulting polyblend. The dispersion mechanism is driven by the capability of a compounder to generate a stress history of sufficient peak magnitude to disrupt the stabilizing interfacial forces of the blend and by the complicated deformations generated during plastication. The melt mixing dispersion process is controlled by the shear viscosity and elasticity ratios of the components and the deformation field generated by the operating condition and mixer geometry. Clearly, a blend can be mixed in a non-optimal fashion whereby agglomerating flow fields and thermodynamic coalescence degrade the dispersion.

The data presented here in no way represent the capability of individual compounders or the particular dispersion problems of polymer blends. Recipes containing crystalline materials, lubricating additives, multicomponents, components with high viscometric function ratios, interfacially reacting components, degradable ingredients, low bulk density powders or agglomerating feedstocks present interesting challenges with real commercial value to the compounding technologist. Equipment capabilities with respect to heat transfer, downstream feeding, multiple injection, feed intake capability, effective venting design and low pressure discharge often dictate candidates for a successful process. Furthermore, the compounder is a part (although the heart) of a manufacturing cell that could include preblenders, feeders, pumps, temperature control systems, vacuum equipment and die and quench systems, which must be coordinated as an operable and optimal system.

Acknowledgements

The mixer study was conducted as an auxiliary program to an IUPAC program called 'Melt Rheology and Concomitant Morphology' in *Polyblends/Alloys*, A.P. Plochocki (coordinator).

References

1. Herrmann, H. (1972) *Schneckenmaschinen in der Verfahrenstechnik*, Springer, Berlin.
2. White, J.L. (1990) *Twin Screw Extrusion, Technology and Principles*, Hanser, New York.
3. Taylor, G.I. (1932) The viscosity of a fluid containing small drops of another fluid. *Proc. R. Soc.* **A138**, 41.
4. Taylor, G.I. (1934) The formation of emulsions in definable fields of flow. *Proc. R. Soc.* **A146**, 501.
5. Van Oene, H.J. (1972) *J. Coll. Interf. Sci.* **40**, 448.
6. Elmerdorp, J.J. and Maalke, R.J. (1986)/(1985) A study on polymer blending microrheology. *Polym. Eng. Sci.* **26**, 415/**25**, 1041.
7. Rumscheidt, F.D. and Mason, S.G. (1961) *J. Colloid Sci.* **16**, 238.
8. Plochocki, A.P. (1982) Development of industrial polyolefin blends using melt rheology data. *Polym. Eng. Sci.* **22**, 1153.
9. Han, C.D. (1981) *Multiphase Flow in Polymer Processing*, Academic Press, Chapters 4 and 5.
10. Plochocki, A.P., Dagli, S.S. and Andrews, R.D. (1990) The interface in binary mixtures. *Polym. Eng. Sci.* **30**, 741.
11. Serpe, G., Jarrin, J. and Dawans, F. (1990) Morphology processing relationships in polyethylene-polyamide blends. *Polym. Eng. Sci.* **30**, 553.
12. Shih, C.K., Tynan, D.G. and Demelsbeck, D.A. (1991) Rheological properties of multicomponent polymer systems undergoing melting or softening during compounding. *Polym. Eng. Sci.* **31**, 16–70.
13. Min, K. (1987) *Interm. Polym. Processing.* **1** 179.
14. Spencer and Wiley (1951) *J. Colloid Sci.* **6**, 133.
15. Gotsis, A.D. *et al.* (1991) *Summary report*, PPI Mixing Study.
16. Tadmor, Z. (1988) *AIChEJ* **34**, 1943.
17. Manas-Zloczower, I., Nir, A. and Tadmor, Z. (1982) Dispersive mixing in internal mixers—a theoretical model based on agglomerate rupture. *Rubber Chem. Technol.* **55**, 1250.
18. Curry, J. and Kiani, A. (1990)/(1991) Measurement of stress level in continuous melt compounders. *SPE Antec Reprints* **48**, 1599.
19. David, B. (1990) PhD Thesis, Technion.
20. Werner, H. (1976) Dr. Ing Thesis, TU Munich.
21. Curry, J., Kiani, A. and Dreiblatt, A. (1991) Feed variance limitations of co-rotating intermeshing extruders. *Int. Poly. Proc. VI* **2**, 148.
22. Plochocki, A.P. (1986) *MIXERIU*. A funded university-corporate technology investigation.
23. Plochocki, A.P. (1992) Melt rheology in concomitant phase structure in a model binary mixture of industrial polymers. *Pure Appl. Chem.* (in prep.).
24. Mack, M.E., Plochocki, A.P. and Dagli, S.S. (1988) Morphologie von Polymergemischen und Polymerblends. *Kunstoffe* **78**, 254.
25. Plochocki, A.P., Dagli, S.S., Curry, J. and Starita, J. (1989) Effect of mixing history on phase morphology of a polyalloy and a polyblend. *Polym. Eng. Sci.* **29**, 617.
26. Karian, H.G. and Plochocki, A.P. (1987) Mixing studies of a polyalloy via co-rotating twin-screw extrusion. *SPE Antec Reprints* **45**, 1334.
27. Valsamis, L.N., Kearney, M.R., Dagli, S.S., Mehta, D.D. and Plochocki, A.P. (1988) Phase morphology of a model polyblend fabricated in industrial mixers. *SPE Antec Reprints* **45**, 1316.

3 Compatibilisation and reactive blending

J.G. BONNER and P.S. HOPE

3.1 Introduction

The achievement of compatibilisation, whether by addition of a third component (a so-called 'compatibiliser') or by inducing *in situ* chemical reaction between blend components (reactive blending), has played an important role in the development of polymer blends. Indeed most commercial blends are considered to be compatible. The first question must therefore be 'what do we mean by compatibility?' The answer is not necessarily straightforward, as many workers in the past have used different definitions.

Compatibility is frequently defined as miscibility on a molecular scale (e.g. [1]). This undoubtedly has the merit of clarity, but has the disadvantage of confining the definition of compatibility to encompass only those blends showing true thermodynamic miscibility, and thereby excluding a very large number of blends, both academically studied and commercialised, which many workers would consider compatible. Another way of defining compatible blends is as polymer mixtures which do not exhibit gross symptoms of phase separation. This widens the scope considerably, and it is certainly true that most compatibilised blends contain very finely dispersed phases, but the definition still excludes some blends which have been modified to facilitate the generation of a preferred, but not necessarily fine, morphology and hence preferred physical properties. One example of such a system would be the compatibilised polyethylene/polyamide blends developed by Du Pont to give blow-moulded containers with enhanced barrier properties to solvents (see [2] for example).

A third definition, preferred by the present authors, is simply to consider blends as compatible when they possess a (preferably commercially) desirable set of properties. This leaves unanswered the question of how this is achieved, and therefore allows the materials developer free rein to exploit any avenue which will lead to a technologically useful product.

The purpose of this chapter is to review compatibilisation technology (sections 3.2 and 3.3), after which the most extensively studied systems will be discussed in detail (sections 3.4 and 3.5). For this purpose a distinction is drawn between blends compatibilised by addition of a third component (compatibiliser), and those produced by reactive blending. Within these headings the blends are described under major groupings; to some extent this has been done arbitrarily, as some blends may clearly fit more than one grouping. Finally, future trends are identified (section 3.6)

The focus of the chapter is on thermoplastic blends produced by melt compounding. For this reason we have chosen to exclude discussion of rubber blends (including thermoplastic elastomers), interpenetrating networks and thermosetting blends.

3.2 Compatibilisation mechanisms

In most cases, melt mixing two polymers results in blends which are weak and brittle; while the low deformation modulus may follow an approximately linear mixing rule, the ultimate properties certainly will not. This is because the incorporation of a dispersed phase in a matrix leads to the presence of stress concentrations and weak interfaces, arising from poor mechanical coupling between phases. It is most common for compatibilisation to be achieved by addition of a third component, or by *in situ* chemical reaction, leading to modification of the polymer interfaces in two-phase blends, and thereby to tailoring of the phase structure, and hence properties.

The factors contributing to end-use properties during manufacture of a blend by melt compounding, and subsequent conversion processing to produce a finished article, are illustrated in Figure 3.1. The mechanical properties of a blend or alloy will be determined not only by the properties of its components, but also by the phase morphology and the interphase adhesion, both of which are important from the viewpoint of stress transfer within the blend in its end-use application. The phase morphology will normally be determined by the processing history to which the blend has been subjected, in which such factors as the process (mixer type, rate of mixing and temperature history), the rheology of the blend components and the interfacial tension between phases in the melt are important. The phase morphology is unlikely to be in thermodynamic equilibrium, but generally will have been stabilised against de-mixing by some method or other; this usually means via quenching to below the glass transition temperature of one or both phases, or via the occurrence of crystallinity in one or both phases, or occasionally by cross-linking.

In any case, it is readily understood that compatibilisation can in principle interact in complex ways to influence final blend properties. One effect of compatibilisers is to reduce the interfacial tension in the melt, causing an emulsifying effect and leading to an extremely fine dispersion of one phase in another. Another major effect is to increase the adhesion at phase boundaries, giving improved stress transfer. A third effect is to stabilise the dispersed phase against growth during annealing, again by modifying the phase-boundary interface. In practice it is likely that all these effects will occur to some extent with addition of a particular compatibiliser, and that the possibility of other effects (such as modification of rheology) may also occur.

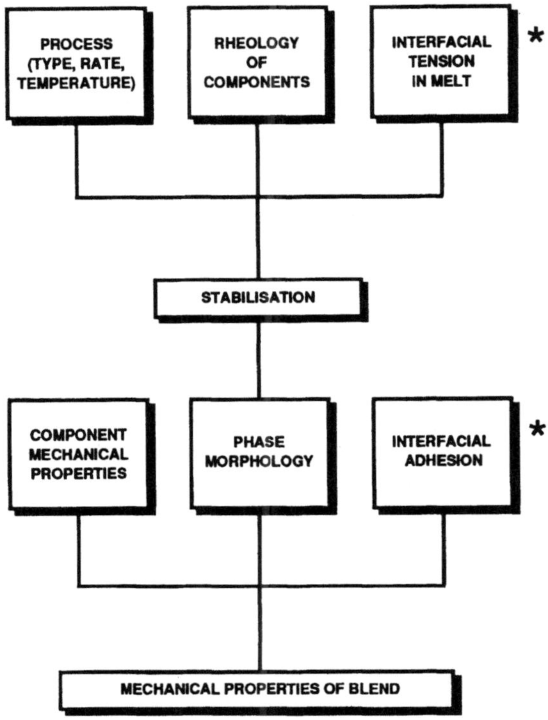

Figure 3.1 Summary of the factors contributing to end-use properties in melt compounded blends, highlighting the role of compatibilisers (*).

This complexity can be illustrated by the findings of experiments aimed at selecting compatibilisers for blends of high density polyethylene (HDPE) with other commercial polymers [3]. A Brabender Plasticorder was used to prepare melt mixtures of HDPE with nylon 6, nylon 6-6, nylon 6-3T (an amorphous polyamide) and polyethylene terephthalate (PET), with and without low levels of various proprietary compatibilising agents. The blends were characterised in terms of phase morphology (by scanning electron microscopy) and by tensile testing of samples cut from moulded plaques. The maximum phase sizes, and the tensile elongation at break results, are summarised as a function of compatibiliser addition level in Table 3.1. It was clear that in many cases compatibiliser addition had a remarkable effect on phase dispersion (see Figure 3.2, for example), and could cause substantial improvements in tensile elongation behaviour. However, achievement of the finest phase dispersion did not in itself guarantee the highest values of ultimate elongation, confirming the complex nature of the compatibilising effect. It was also suggested in this work that the method of compatibiliser selection should reflect the end-use requirements of the blend.

Table 3.1. Comparison of maximum phase size (d_{max}) and average tensile elongation at break (ε_B) results for HDPE blends containing 15% by weight of various polymers and low levels of proprietary compatibilisers (after [3])

Compatibiliser level (pph)	Nylon 6		Nylon 6-6		Nylon 6-3T		PET	
	d_{max} (μm)	ε_B (%)	d_{max} (μm)	ε_B (%)	d_{max} (μm)	ε_B (%)	d_{max} (μm)	ε_B (%)
0	60	20	100	10	15	220	65	10
0.1	5	40	13	15	6	600	15	320
2	5	130	2	65	10	740	20	>800
5	5	110	6	65	4	300	20	400

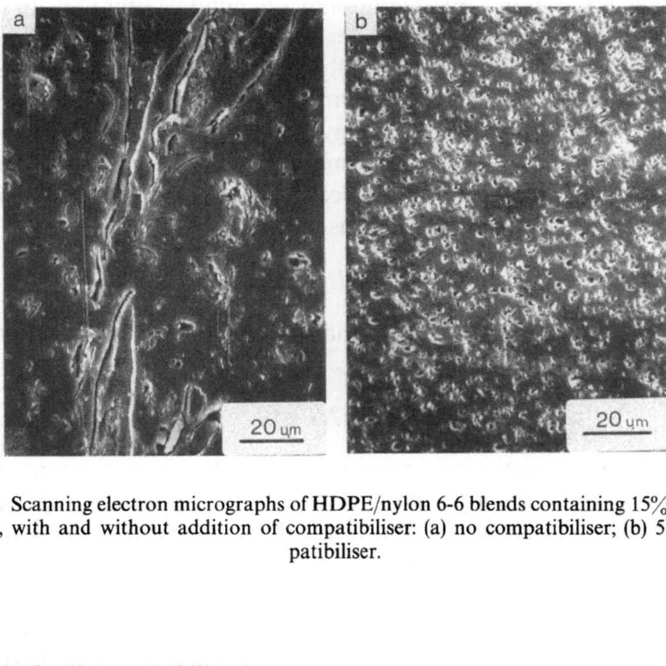

Figure 3.2 Scanning electron micrographs of HDPE/nylon 6-6 blends containing 15% by weight nylon 6-6, with and without addition of compatibiliser: (a) no compatibiliser; (b) 5 pph compatibiliser.

3.3 Methods of compatibilisation

Pursuing our technological definition of compatibilisation as modification of blends to produce a desirable set of properties, it is perhaps unhelpful to attempt to categorise methods too narrowly. However, a number of different lines of approach can be defined which may assist the materials developer. Broadly, these are:

1. Achievement of thermodynamic miscibility.
2. Addition of block and graft copolymers.

3. Addition of functional/reactive polymers.
4. *In situ* grafting/polymerisation (reactive blending).

There is clearly overlap between these approaches, and which to use will depend on the requirements of the situation. For example, polymer manufacturers may have the scope to tailor their materials during production to facilitate compatibilisation, whereas producers of technological compounds, or polymer converters, are likely to be restricted to the use of commercially available polymers. Clearly economic factors will also play a substantial part in deciding how to proceed.

3.3.1 Thermodynamic miscibility

Compatibilisation by the achievement of true thermodynamic miscibility is a concept which has been exploited in only a handful of situations to produce commercial blends, the most striking example being blends of polyphenylene oxide (or polyphenylene ether) with polystyrenes. Briefly, it is recognised that miscibility between polymers is determined by a balance of enthalpic and entropic contributions to the free energy of mixing. While for small molecules the entropy is high enough to ensure miscibility, for polymers the entropy is almost zero, causing enthalpy to be decisive in determining miscibility. The change in free energy on mixing (ΔG) is written as

$$\Delta G = \Delta H - T\Delta S$$

Where H is enthalpy, S is entropy and T is temperature. For spontaneous mixing, ΔG must be negative, and so

$$\Delta H - T\Delta S < 0.$$

This implies that exothermic mixtures ($\Delta H < 0$) will mix spontaneously, whereas for endothermic mixtures miscibility will only occur at high temperatures. For two-component blends it is possible to construct a phase diagram, which may exhibit lower or upper critical solution temperature (LCST or UCST) (Figure 3.3). In practice, LCST behaviour is more commonly seen, phase separation occurring as temperature increases, because the intermolecular attractive forces responsible for the miscible behaviour tend to disappear as the internal energy of the molecules becomes high enough to overcome them.

In principle it should be possible to tailor the structure of polymers in some circumstances to modify their phase diagram in blends, and thereby to achieve miscible blends. In practice the development of miscible blends has been empirical, and the state of the art is such that prediction of miscibility is arguably not a realistic option currently available to the materials developer. However, it is interesting to note that examples exist in the literature where addition of a third component to a blend has caused a phase change from a two-phase to a single-phase system.

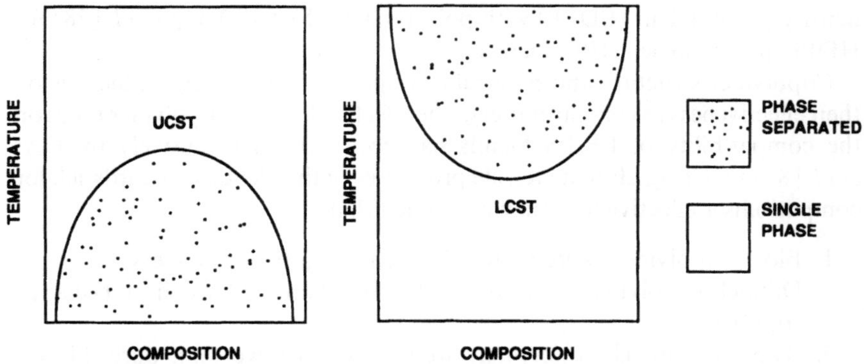

Figure 3.3 Schematic phase diagrams for binary blends showing LCST and UCST behaviour.

3.3.2 Addition of block and graft copolymers

The addition of block or graft copolymers represents the most extensively researched approach to compatibilisation of blends. Block copolymers have been more frequently investigated than graft copolymers, and in particular block copolymers containing blocks chemically identical to the blend component polymers. It is perhaps not surprising that block and graft copolymers containing segments chemically identical to the blend components are obvious choices as compatibilisers, given that miscibility between the copolymer segments and the corresponding blend component is assured, provided the copolymer meets certain structural and molecular weight requirements, and that the copolymer locates preferentially at the blend interfaces. The classical view of how such copolymers locate at interfaces is shown in Figure 3.4, and experimental verification that this happens has been found for many systems; for example Barentsen *et al.* [4] and Fayt *et al.* [5, 6], both for blends of low

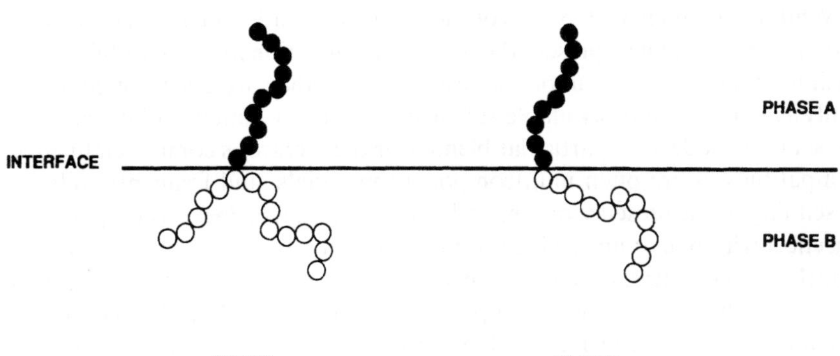

Figure 3.4 Schematic diagram showing location of block and graft copolymers at phase interfaces.

density polyethylene (LDPE) with polystyrene (PS), and Yang *et al.* [28] for HDPE/polypropylene (PP) blends.

Copolymer structure and molecular weight have important influences on their effectiveness as compatibilisers. The effect of different copolymer types on the compatibility of PE/PS blends has been studied extensively by Fayt *et al.* [8–11]. Using ultimate tensile properties of the blends as a yardstick for compatibiliser effectiveness, they concluded that:

1. Block copolymers were more effective than graft copolymers.
2. Diblock copolymers were more effective than triblock or star-shaped copolymers.
3. 'Tapered' diblock copolymers were more effective than pure diblock copolymers.

As far as molecular weight is concerned, Paul [12] suggests that solubilisation of a discretely dispersed homopolymer into its corresponding domain of a block copolymer compatibiliser only occurs when the homopolymer molecular weight is equal to or less than that of the corresponding block. However, stabilisation of a matrix homopolymer into its corresponding domain of a block copolymer compatibiliser will occur even if the molecular weights are mismatched. Gaylord [13] offers the pragmatic view that a balanced molecular weight is needed for copolymer compatibilisers; the segments need to be long enough to anchor to the homopolymer (i.e. to solubilise) but short enough to minimise the amount of compatibiliser needed, and hence to be cost-effective.

The requirement that the copolymer should locate preferentially at the blend interfaces also has implications for the molecular weight of the compatibiliser. Both the thermodynamic 'driving force' to the interface and the kinetic 'resistive force' to diffusion increase with molecular weight, suggesting that high molecular weight copolymers may be used if sufficiently long times are available during the process, but that lower molecular weights must be used if available diffusion times are short.

While copolymers with blocks of chemical composition identical to those of the two homopolymers present the best technical option for compatibilisation, such materials suffer a number of disadvantages; they are often not commercially available, or only available at high cost, and may limit flexibility, needing to be tailor-made for a particular blend. For these reasons commercially used compatibilisers are often multicomponent materials, and frequently rely on miscibility or the presence of some other interaction to achieve a compatibilising effect with one or more blend components. For example, ethylene–propylene (EP) copolymers, which contain a large number of randomly polymerised units, in addition to ethylene and propylene blocks, have found some success as compatibilisers for blends of PE and PP (e.g. [14–16]). Another example is the use of commercially available styrene–ethylene/butene–styrene (SEBS) triblock copolymers, which have proved effective in compatibilising blends of HDPE with PET.

3.3.3 *Addition of functional polymers*

The addition of functional polymers as compatibilisers has been described by many workers. Usually a polymer chemically identical to one of the blend components is modified to contain functional (or reactive) units, which have some affinity for the second blend component; this affinity is usually the ability to chemically react with the second blend component, but other types of interaction (e.g. ionic) are possible. The functional modification may be achieved in a reactor or via an extrusion-modification process. Examples include the grafting of maleic anhydride or similar compounds to polyolefins, the resulting pendant carboxyl group having the ability to form a chemical linkage with polyamides via their terminal amino groups. Functionalised polymers (usually maleic anhydride or acrylic acid grafted polyolefins) are commercially available at acceptable cost to be used as compatibilisers.

3.3.4 *Reactive blending*

A comparatively new method of producing compatible thermoplastics blends is via reactive blending, which relies on the *in situ* formation of copolymers or interacting polymers. This differs from other compatibilisation routes in that the blend components themselves are either chosen or modified so that reaction occurs during melt blending, with no need for addition of a separate compatibiliser. This route has found commercial application, for example in blends of polycarbonate and polyesters, and toughened polyamides, which are blends of polyamides with graft-functional polyolefin elastomers. Graft-functionalised elastomers, produced by melt modification, are also commercially available for toughening nylons.

Although batch-type melt mixers may be used for reactive blending, continuous processing equipment such as single- and twin-screw extruders are often preferred. As well as the advantages of continuous production, these units generally have better temperature control, and can be designed to allow for removal of unwanted reaction products by devolatilisation.

A number of reactive blending mechanisms may be exploited:

1. Formation *in situ* of graft or block copolymer by chemical bonding reactions between reactive groups on component polymers; this may also be stimulated, for example, by addition of a free radical initiator during blending.
2. Formation of a block copolymer by an interchange reaction in the backbone bonds of the components; this is most likely in condensation polymers.
3. Mechanical scission and recombination of component polymers to form graft or block copolymers. This is generally induced by high shear levels during processing.
4. Promotion of reaction by catalysis.

The area of reactive blending is one in which there is currently a great deal of development activity, and much proprietary knowledge.

3.4 Systems using compatibiliser addition

3.4.1 Polyethylene/polystyrene blends

Compatibilisation of PE/PS blends can readily be achieved by addition of graft or block copolymers. This system is among the earliest and most extensively studied, with notable contributions from Heikens and coworkers, Paul and coworkers and Teyssie and coworkers.

Barentsen and Heikens [20] and Barentsen *et al.* [4] describe the addition of graft copolymers of LDPE with PS (PS-g-LDPE) to LDPE/PS blends. The graft copolymers were prepared by Friedel-Crafts alkylation of the aromatic rings in PS with the olefinic groups in LDPE, using an $AlCl_3$ catalyst, and contained practically equal weight fractions of ethylene and styrene. Blends were produced by melt mixing at 195 °C on a laboratory mill, and the graft copolymer was first melt blended with the polymer forming the dispersed phase before being added to the matrix polymer. Additon of 7.5% by weight copolymer caused a substantial reduction in size of the dispersed phase, both for PS-rich and LDPE-rich blends. It was also clear from scanning electron micrographs of fracture surfaces of the blends that the copolymer concentrated at the interface between the LDPE and PS phases, and adhered to both phases (Figure 3.5). This was confirmed by peel test measurements. The blends

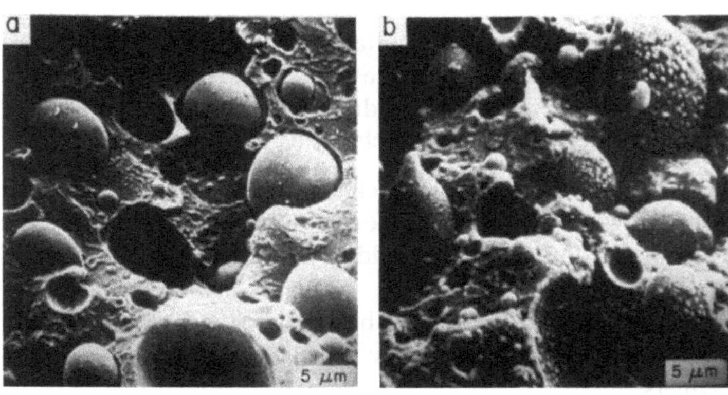

Figure 3.5 Scanning electron micrographs of fracture surfaces of PS/LDPE blends with and without graft copolymer compatibiliser: (a) 75/25 PS/LDPE; (b) 75/23.75/1.25 PS/LDPE/graft (from Barentsen, Heikens and Piet, 1974, by permission of the publishers, Butterworth Heinemann Ltd. ©).

containing copolymer also possessed higher yield strength, elongation and breaking strength than the unmodified blends, across the whole composition range, as well as having significantly increased impact strength.

In a similar series of experiments Barentsen *et al.* [21] added a graft copolymer of PS and an amorphous EP copolymer (PS-g-EP) to EP/PS blends, again using a mill. It was found that adding the amorphous PS-g-EP to PS-rich EP/PS blends enhanced the impact strength much more than PS-g-LDPE did in the corresponding LDPE/PS blends, while causing a similar reduction in dispersed phase size, suggesting that deformation and fracture mechanisms are strongly influenced by the crystallinity of the dispersed phase.

The use of PS-g-LDPE graft copolymers to compatibilise LDPE/PS blends has also been extensively investigated by Locke and Paul [22, 23], who formed the graft copolymer by radiation grafting styrene to LDPE. LDPE pellets swollen with styrene monomer were exposed to cobalt 60 γ-radiation, and the copolymer produced from varying radiation doses was extracted and characterised. LDPE/PS (50/50) blends containing various levels of graft copolymer (up to 33% by weight) were produced by melt blending in a Brabender Plasticorder (10 min at 170 °C). Substantial improvements in yield strength and elongation were reported on adding copolymer, and phase size was reduced at all addition levels. The most effective copolymer in terms of improving strength and elongation was that produced at a radiation dose of 0.5 megarads, which was found to contain approximately 50% by weight PE, although in this case the authors suggest the overriding factor in controlling effectiveness of the compatibiliser was the competition between the extent of grafting and crosslinking occurring during irradiation. The mechanism of blend property improvement was believed to be the increased interfacial adhesion provided by the graft copolymer.

Heikens *et al.* [17] studied the effect of adding small amounts of both graft and block copolymers of PS and LDPE, with a wide range of structures, on the morphology and mechanical properties of PS-rich LDPE/PS blends. The detailed fine structure of the copolymers gave rise to large effects, not only on impact strength but also on the magnitude of the tensile modulus of the blends. It was inferred that block copolymers had a tendency to form a third, low modulus, dispersed phase, in addition to being present at the LDPE/PS interface, effectively preventing some of the PS in the copolymer from contributing to modulus.

Extensive studies of compatibilisation of LDPE/PS blends using block copolymers were carried out by Fayt *et al.* [5, 6, 8–10]. These workers melt blended LDPE and PS in a two roll mill at 210 °C in the presence of 9% by weight diblock styrene–ethylene butadiene (S-EB) copolymers, produced by a two-step anionic polymerisation process and subsequently hydrogenated using Ziegler-type catalysts. The blends containing block copolymer showed greatly improved dispersion of the minor phase, which was retained after compression moulding, the LDPE-rich blends containing PS particles smaller than 1 μm while PS-rich blends showed a fine, semicontinuous or continuous two-phase structure. Both strength and ductility of the blends were greatly improved by addition of copolymer, particularly for PS-rich blends, and larger values of energy to break were reported. In particular it was shown that the block copolymers were superior to the previously-discussed graft copolymers as emulsifiers, and produced a better balance of properties in the blends over a wide composition range (Table 3.2).

The same authors compared the performance of the S-EB diblock copolymer with that of a 'tapered' diblock copolymer (effectively a PS–PS/PB random-hydrogenated PB triblock copolymer) of the same composition and molecular weight. The tapered diblock was found to be more effective than the pure diblock copolymer as a compatibiliser, in that it yielded blends with finer morphology (for LDPE-rich blends) and superior mechanical properties. The authors speculate that the tapered copolymer could be providing a graded modulus at the interphase boundary, resulting in improved mechanical response. The location of block copolymer at the LDPE/PS interface was confirmed by transmission electron microscopy on blends containing S-EB copolymer modified by addition of a central polyisoprene block, which provided means of selective staining. A rather regular interface of approximate thickness 100 Å was observed.

S-EB diblock copolymers have also been added to HDPE/PS blends with similar results [8,9]. Addition of copolymer again greatly reduced and stabilised the size of the dispersed phase, and tensile mechanical properties were improved across the composition range. Ductility was improved sufficiently to follow an approximately additive mixing rule for elongation at break, and strength was again markedly improved for PS-rich blends; this was attributed to improved interphase adhesion, and to the interlocking morphology seen at high PS concentrations.

Table 3.2. Maximum improvements in ultimate tensile strength and elongation at break for LDPE/PS blends modified by graft and block copolymers (after [8,9])

LDPE/PS composition (by weight)	Copolymer type	% Copolymer added (by weight)	Improvement in ultimate tensile strength (%)	Improvement in elongation at break (%)	Reference
70/30	Graft	9	100	*	[20]
	Block	9	70	900	[8,9]
	Block	9	90	3000	[8,9]
50/50	Graft	33	60	120	[22,23]
30/70	Graft	9	80	*	[20]
	Block	9	120	120	[8,9]
	Block	9	60	1300	[8,9]

* Reduction in elongation at break

Lindsay *et al.* [40] demonstrated that an SEBS triblock copolymer when added to HDPE/PS greatly improved the ductility of the blend, but with an accompanying loss in strength and modulus. The SEBS triblock copolymer was a commercial resin, Kraton G-1652, from Shell Chemical Company, and the blends were manufactured using a laboratory single-screw extruder and

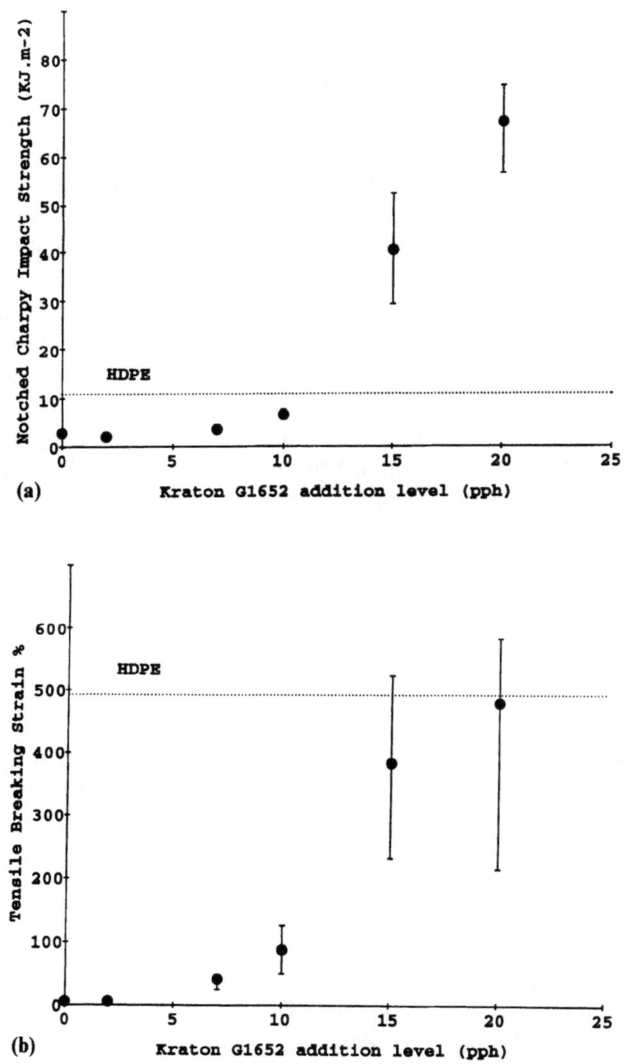

Figure 3.6 Effect of addition of S-EB triblock copolymer on (a) impact and (b) ductility of a compression moulded HDPE/HIPS blend containing 20% by weight HIPS (from Miles *et al.* [25]).

subsequently injection moulded. Copolymer addition levels between 5 and 30% by weight were studied, and gave substantial increases in tensile elongation at break over the whole range of HDPE/PS blend compositions. The particular value of improving the toughness of PS-rich blends was stressed.

The effectiveness of SEBS triblock copolymers as compatibilisers for blends of HDPE with high impact polystyrene (HIPS) has been confirmed by Miles *et al.* [25]. Using blends produced on a twin-screw extruder it was shown that 10–15 pph copolymer addition was sufficient to fully recover the impact and tensile elongation performance of a blow-moulding grade HDPE containing 20% by weight HIPS (Figure 3.6).

Also in more recent work Fayt *et al.* [5, 6, 11] have studied more systematically the effect of hydrogenated (PB-b-PS) on blends of PS with LDPE, LLDPE, HDPE and hydrogenated PB. They concluded that in all cases phase size was significantly reduced and stabilised against coalescence in further processing, and interfacial adhesion was dramatically increased. Diblock copolymers with balanced composition were shown to be the most efficient interfacial agents, so much so that only 1–2% by weight would provide homogeneous and stable dispersions. The mechanical performance of the blends was also shown to be strikingly dependent on the molecular characteristics of the copolymers, diblocks again being more effective than graft, triblock or star-shaped copolymers in improving elongation and tensile strength. By using a copolymer of optimised structure (pure diblock, high molecular weight) at 10% by weight loading to modify a PS/LDPE blend containing 20% by weight LDPE, it was possible to achieve toughening equivalent to that found in commercial HIPS. It has also been shown [26] that the rheological properties of HDPE/PS blends compatibilised using hydrogenated SB diblock copolymers are very sensitive to the level of compatibiliser addition, particularly in the low frequency region.

3.4.2 Polyolefin blends

Among the largest classes of polyolefin blends in use today are blends of isotactic PP and ethylene–propylene rubbers (EPR), which range in properties from rubber-toughened PP to thermoplastic elastomers. PP has also been toughened by blending with HDPE or LLDPE, especially with a view to improving low temperature impact performance.

Nolley *et al.* [14] showed that EP copolymers could serve as compatibilising agents for PP/LDPE blends. Those copolymers, which displayed residual crystallinity because of longer ethylene sequences, were more effective than purely amorphous copolymers. Ho and Salovay [15] and Bartlett *et al.* [16] have shown that addition of 5% EPR is necessary to ensure a linear relationship between tensile strength and composition for HDPE/PP blends. Dumoulin *et al.* [27] investigated the addition of EP block copolymers to LLDPE/PP blends, which resulted in improved modulus and strength (especially at −40 °C), and reduced elongation; however, the effects were fairly marginal, and the compatibiliser is arguably not needed for this system.

Yang *et al.* [28] describe a complex picture when toughening PP by blending with HDPE and EPDM. These workers report that replacing up to 20% HDPE (in a 20/80 HDPE/PP blend) by EPDM led to improved impact and tensile properties. In this case the EPDM appeared to act both as impact modifier and as compatibiliser; scanning electron microscopy evidence suggests that the EPDM existed in these blends both as discrete particles and as an interlayer between the HDPE and the PP. Addition of EPDM also affected the spherulitic morphology of the blend.

In recent work Lohse *et al.* [29] have reported useful compatibilisation of EP/PP blends by addition of up to 10% by weight EP-g-PP graft copolymers, leading to a four-fold reduction in phase size and doubling of low temperature impact strength. The EP-g-PP graft copolymer was synthesised by growing pendent PP grafts from the unreacted double bonds in an EPDM, using a two-stage polymerisation process.

More recently Miles *et al.* [25] have explored the potential to compatibilise blends of PP with LDPE, LLDPE, and HDPE in order to recover the mechanical properties of recycled polyolefins. Using 80/20 PE/PP blends produced on a twin-screw extruder, a wide range of commercially available polymers was screened for their effectiveness as compatibilisers. In summary, the impact and ultimate tensile elongation of LLDPE/PP blends could readily be recovered by addition of 2 pph of an EP random copolymer, or a very low density polyethylene (VLDPE). The homogeneity of the blends was also substantially increased (see Figure 3.7, for example). Blends of LDPE or

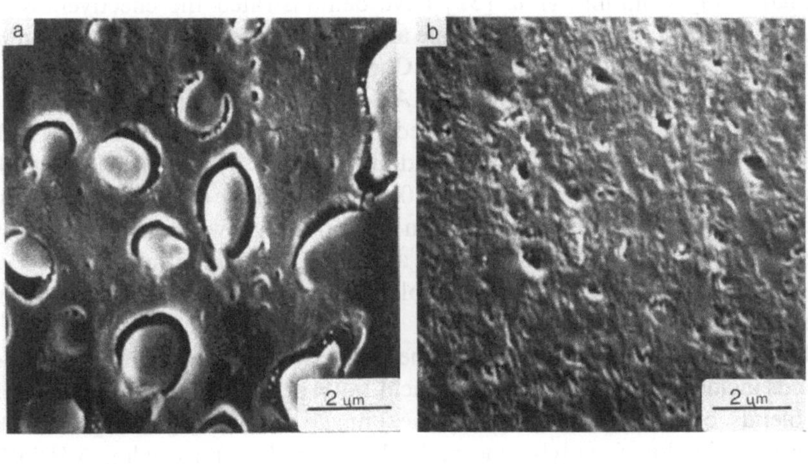

Figure 3.7 Scanning electron micrographs of LLDPE/PP blends containing 20% by weight PP with and without addition of a compatibiliser: (a) no compatibiliser; (b) with 2 pph EP copolymer.

HDPE with PP were more difficult to compatibilise, although some improvement could be obtained on addition of EP random copolymer, provided higher levels were used.

3.4.3 *Blends containing polyamides*

Blending of polyamides is carried out commercially with a number of objectives, the main being to increase their toughness (by incorporation of a dispersed rubbery phase). Other uses include imparting lubricity to polyphenylene sulphide and improvement of hydrocarbon and other solvent barrier performance for polyethylene containers.

A commonly used method of inducing compatibility between polyamides and polyolefins is by chemical modification of the polyolefin to contain pendent carboxyl groups, often by grafting with maleic anhydride (MAH) or similar compounds, which form chemical linkages to the polyamide via the terminal amino groups. While most commonly exploited in reactive blending, the concept has been employed to produce compatibilisers by grafting maleic anhydride onto the polyolefin, and then adding the MAH-g-polyolefin as a third component in polyamide/polyolefin blends. An early example is work by Ide and Hasegawa [30], who added various levels of MAH-g-PP to extrusion compounded blends of PP with nylon 6. Addition of MAH-g-PP caused a dramatic reduction in the phase size of blends containing 20% by weight nylon 6, accompanied by remarkable improvements in tensile strength, elongation and impact strength. These improvements were also seen for nylon 6-rich blends. The existence of a reaction between the MAH and the nylon 6 amino end-groups was established by solvent extraction of the PP phase, estimation of the number of amino end-groups and DSC analysis of the residue. In a similar study Cimmino *et al.* [31] have demonstrated the effectiveness of MAH-g-EPDM in producing a stable and more uniform phase dispersion in EPDM/nylon 6 blends. More recently Chen *et al.* [32] investigated the effects of adding a commercially available MAH-g-PP compatibiliser on the phase morphology of extrusion compounded blends of HDPE and LDPE with nylon 6 and nylon 11. In all cases addition of 5% by weight MAH-g-PP resulted in a substantial reduction ($\times$ 10) in the size of the dispersed phase, but in addition caused a dramatic reduction in the phase growth seen in the binary blends on annealing for up to 1.5 hours at 230 °C. The compatibilising effect of MAH-g-PE in polyethylene/nylon 6 blends has also been reported by Kim *et al.* [33].

Other approaches to enhancing the properties of polyolefin/polyamide blends include the addition of acrylic acid/butyl acrylate/styrene terpolymers to blends of nylons and polyethylene impact modifiers [34, 35] and the addition of nylon 6–polybutene multiblock copolymer to nylon 6/HDPE blends [36]. In the work by Lindberg [36] the presence of a strong interaction between the blend phases was detected by melt viscosity measurements, and although elongation at break for the blends was improved, other mechanical properties were unchanged.

Ide and Hasegawa [30] have also investigated compatibilisation of nylon 6/polystyrene blends by addition of a styrene-methacrylic acid (MAA) copolymer. The authors claim improved dispersibility for the compatibilised blend, caused by the same reaction, between terminal amino groups in nylon 6 and the MAA in the copolymer, as seen in nylon 6/PP blends compatibilised using MAH-g-PP. Although pellet appearance was improved on adding compatibiliser, no significant improvements in physical properties of the blend were seen. Chen *et al.* [32] found incorporation of styrene-maleic anhydride (SMA) copolymer much more effective than styrene acrylonitrile (SAN) copolymer in compatibilising nylon 6/polystyrene blends; incorporation of 5% SMA caused more than ten-fold reduction in the size of the dispersed phase, compared to two-fold for SAN, over a wide range of nylon 6/polystyrene compositions. SMA was also more effective at stabilising the blend structure against phase growth during annealing. However, interesting results were found by Angola *et al.* [37], who demonstrated that SMA copolymers had a strong compatibilising effect in blends of nylon 6 with SAN copolymers containing 70% by weight nylon 6. Addition of SMA caused a finer dispersion of SAN in nylon 6, and improved the tensile strength and impact properties of the blend. The authors suggest this was due to formation *in situ* of nylon-SMA graft copolymer, which provided improved adhesion and stress-transfer at the dispersed particle interface. The incorporation of a finely dispersed brittle polymer (SAN) in the ductile nylon 6 matrix was claimed as a new way of obtaining toughened plastics.

3.4.4 *Blends containing polyethylene*

In addition to the blends with PS and polyolefins already described, PE has been studied in blends with a wide range of polymers. Because PE is essentially non-polar, these studies have inevitably involved compatibilisation in some form.

Blends of PE with polyvinylchloride (PVC) have been studied by Schramm and Blanchard [38], Wollrab *et al.* [39] and Paul *et al.* [18], among others. Wollrab *et al.* [39] used a graft copolymer of ethylene and vinyl chloride as compatibiliser. Paul *et al.* [18] produced a chlorinated-polyethylene by chlorination in a solid-state slurry, so that the amorphous regions of the PE only were chlorinated, producing a kind of block copolymer. When added to HDPE/PVC or LDPE/PVC blends produced in a Brabender Plasticorder the blends displayed higher ductility, elongation and toughness, but reduced modulus and tensile strength, and the domain size was considerably reduced. CPE containing 36% chlorine was found to be most effective.

Research by Lindsay *et al.* [40] and Traugott *et al.* [41] investigated the use of compatibilisers for blends of HDPE with PET, the impetus being recycling of scrap from beverage bottles. The compatibilisers studied were Epcar 847, an ethylene–propylene–ethylidene norbornene copolymer with crystallisable ethylene blocks, and Kraton G 1652, an SEBS triblock copolymer having styrene end blocks comprising 30% of the total mass and a hydrogenated butadiene midblock. Adding up to 20% by weight SEBS copolymer to a 50/50 HDPE/PET blend in an extruder at 300 °C caused some reduction in modulus and yield strength, but a dramatic increase in ductility, which increased from 3% elongation to greater than 200% elongation (Figure 3.8). The impact strength was also improved. Addition of 20% Epcar had little effect, however. Examination of fracture surfaces and adhesion tests suggested that use of SEBS copolymer caused the blend to form an interpenetrating network of the two immiscible components, and that the copolymer adhered better to the PET and to the HDPE than they did to each other. The authors suggested that interfacial energy considerations led to co-continuity being the preferred state for the blend containing SEBS copolymer, which resided at the interfaces. Both the location of the copolymer at the interface, and the interpenetrating phase structure, enhanced stress transfer and hence elongation and impact properties.

The use of SEBS copolymers has also been investigated for blends of PE with polycarbonate (PC). Endo *et al.* [42] showed that copolymer addition reduced the phase size when PE was the dispersed phase, and prevented phase growth on annealing (260 °C, 90 min). It was suggested that the PE

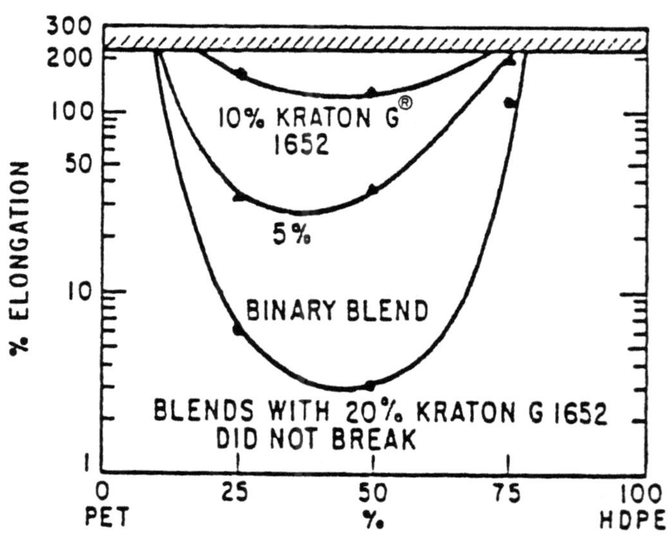

Figure 3.8 Effect of addition of S-EB triblock copolymer on ductility of HDPE/PET blends (from Traugott *et al.*, in *Journal of Applied Polymer Science*, reprinted by permission of John Wiley & Sons, Inc., copyright 1983 ©).

segment of the copolymer dissolved in the PE phase, while the PS segment formed an association with the phenyl rings in the PC.

3.4.5 Blends containing polystyrene

In addition to blends with PE, PS has been studied in blends with a wide range of polymers, the main motive being to toughen the PS. Early work concentrated on blends of PS with rubbers, in an attempt to reproduce HIPS. For example, Kato [43] produced solution blends of styrene with butadiene or isoprene, and Fiese *et al.* [44] investigated the addition of low molecular weight resins to PS/polybutadiene mixtures. The latter authors found that addition of up to 5% low molecular weight resins (Dammor, Cresolnovolak, Coumarone, Colophonium) to blends containing 10% by weight polybutadiene gave substantial improvements in impact strength. The work was interesting in that adhesion tests (between PS and the rubber/compatibilising resin mixture) were successfully used to pre-select compatibilisers. More recently Noolandi and Hong [45] have produced a comprehensive theory for the interface region formed when styrene–butadiene diblock copolymers are added to PS/polybutadiene blends, and Marie *et al.* [46] have studied addition of SB block copolymers to PS/PB blends experimentally. They found that 20% by weight copolymer addition reduced the cloud point by around 60 °C, and that the mixing temperature decreased linearly with copolymer addition. Reiss *et al.* [47] and Kohler *et al.* [48] showed that addition of a styrene–isoprene block copolymer could yield a transparent region in PS/polyisoprene solution blends; Inoue *et al.* [49] took this work further via a systematic study of block copolymer structure on blend microstructure.

Several workers have investigated PS/PP blends. Bartlett *et al.* [50], motivated by the need to improve the properties of scrap PP/PS blends, showed that SEBS triblock copolymer was an effective compatibiliser in injection moulded and compression moulded blends. The effects of adding copolymer were analogous to those described earlier for addition to PE/PS blends, namely increased elongation and impact performance but reduced modulus and tensile strength. For a 50/50 PP/PS blend, adding 20% copolymer caused a ten-fold increase in elongation and a five-fold increase in Izod impact performance. Del Guidice *et al.* [51] investigated addition of a styrene-propylene diblock copolymer, obtained by sequential polymerisation using a Ziegler–Natta catalyst, to PS/PP solution blends. Addition of as little as 5% by weight diblock copolymer dramatically reduced the grain size in blend morphology, and addition of 20% copolymer led to impact strengths greater than those predicted from a linear combination of blend component data.

Finally, an interesting route to compatibilisation was demonstrated by Robeson *et al.* [52]; phase separated blends of PS with poly(α-methylstyrene) were shown to become miscible on addition of a PS-P(αMS) block copolymer, by a compositionally-induced phase change.

3.5 Systems using reactive blending

3.5.1 Polycarbonate/polyester reactive blends

A widely available blend system which relies on reaction to improve compatibility is a polycarbonate/polyester combination. There are many commercially available materials including Xenoy (General Electric), Makroblend (Bayer) and Ultrablend (BASF). The toughness and chemical resistance of polycarbonate can be inadequate for some applications. In order to improve these properties, and therefore to exploit the other excellent properties of polycarbonate in a broader range of engineering-type polymer applications, a thermoplastic polyester is blended with it. A transesterification reaction is thought to occur, producing a compatible blend. Devaux *et al.* [53] have studied this reaction in a blend of polybutylene terephthalate (PBT) and polycarbonate, and concluded that the formation of a four component system occurs. The structures of the four components produced from the esterification reaction were elucidated using infrared and nuclear magnetic resonance spectroscopy, and the average chain lengths of these components were derived together with the nature of the copolyesters produced. It was found that with increasing reaction time the copolyesters changed from block to random copolymers. Kinetic and thermodynamic studies were also made which indicated that the most probable mechanism was a direct, reversible ester–ester reaction. This reaction could be catalysed by titanium residues present in commercial PBT. Figure 3.9 outlines the reaction which was proposed to take place between PBT and PC.

Other workers have studied transesterification reactions in polymer mixtures, in particular polyester/polycarbonate blends. For example, Robeson [54] studied polyarylate/polycarbonate (50/50) blends, which were prepared by melt extrusion at 270 °C and subsequently compression moulded at various

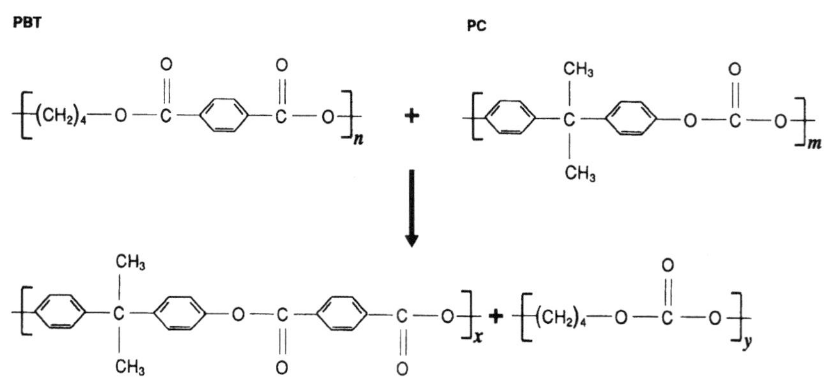

Figure 3.9 Transesterification reaction proposed for PBT/PC blends.

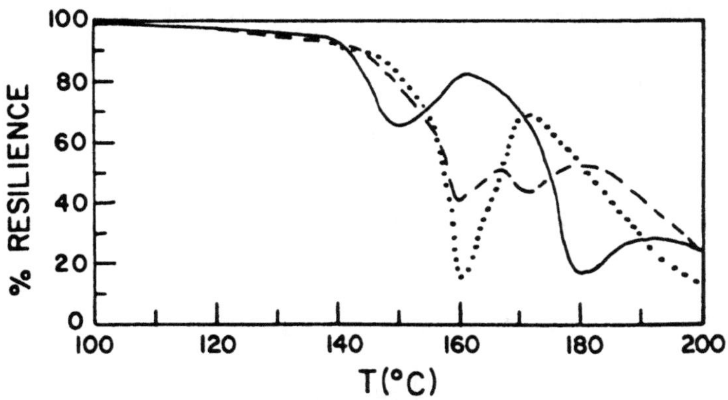

Figure 3.10 Resilience–temperature measurements showing glass transition behaviour of 50/50 polyarylate/polycarbonate blends moulded at different temperatures: (——) 260 °C; (– – –) 320 °C; (-----) 350 °C (from Robeson, in *Journal of Applied Polymer Science*, reprinted by permission of John Wiley & Sons, Inc., copyright 1985 ©).

temperatures (260–350 °C). Compatibility was observed to improve with increased moulding temperature. Figure 3.10 shows results of glass transition temperature (T_g) measurements. With a moulding temperature of 260 °C two values of T_g were observed, similar to those of the two component materials. When moulded at 320 °C some compatibility was evident as the two values of T_g were intermediate between those of the constituent materials. However, when moulded at 350 °C the blend was considered to be homogeneous and totally miscible, displaying a single T_g.

3.5.2 Polyamide reactive blends

Polyamide (PA) has been used extensively in blend systems, many of which rely on a reaction of some kind to provide compatibilisation. Such a blend system which is commercially produced is PA/anhydride modified ethylene–propylene rubber (EPR). Applications for polyamides can be extended by improving the toughness and in particular the impact resistance of the material. Addition of a rubbery polymer has been an effective route to accomplishing this improved toughness, the degree of which is related to the size, shape, volume fraction and interparticle spacing together with the degree of interaction between the rubber phase and the polymer matrix (see, for example, [55]). EPR is not inherently compatible with PA and therefore in order to achieve the desired toughness the most straightforward route would be to add an EPR which was compatible with PA as opposed to adding a third compatibiliser component. This is achieved by modifying the EPR chemically. An acid or anhydride can be grafted to EPR using free radical initiators in melt processing equipment such as an extruder. A maleic anhydride grafted EPR,

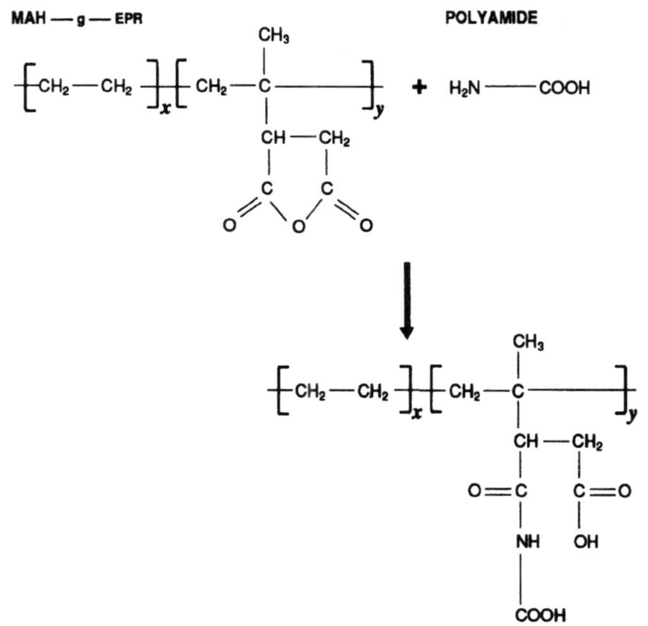

Figure 3.11 Imidisation reaction proposed between maleic anhydride and polyamide.

for example, can then undergo an imidisation reaction with the polyamide (see Figure 3.11).

There are many commercially available rubber toughened polyamides, for example Zytel and Bexloy (DuPont), Nydur (Bayer/Mobay) and Durethan (BASF). In addition maleic anhydride modified EPRs are commercially available, for example Exxelor from Exxon, which material users and compounders can use to produce their own toughened polyamides.

The use of alternative rubbers to EPR has been investigated in polyamide systems, to produce blends with superior performance. Bonner *et al.* [56] have investigated the reactive blending of maleic anhydride grafted nitrile rubber and polyamide 6. Nitrile rubber provides advantages such as good low temperature performance and excellent oil/hydrocarbon resistance, both beneficial properties, for example, in automotive applications. Maleic anhydride grafted nitrile rubber (MAH-g-NBR) was produced by melt mixing MAH with NBR in the presence of a free radical initiator. The grafted material was blended with nylon 6 in a twin screw extruder at an addition level of 10 pph. Unmodified NBR and a commercially available EPR-based polyamide impact modifier were also blended with nylon 6. Test specimens of these blends were injection moulded for Charpy and Izod impact testing at 23 °C and − 20 °C. The tests carried out at 23 °C indicated an improvement in the impact performance of the nylon 6 with the addition of the unmodified NBR.

The impact performance of the nylon 6 was further enhanced when the NBR was replaced by MAH-g-NBR, giving a similar impact strength to the blend containing the commercial additive. At low temperature (– 20 °C), however, the impact performance of the blend containing the commercial additive was substantially reduced, whereas the blend containing the MAH-g-NBR retained its room temperature performance. It was speculated that this may be due to the presence of a high level of grafted MAH coupled with a relatively high degree of crosslinking in the MAH-g-NBR. Figure 3.12 shows the results of the Izod impact testing on the blends.

Kim and Park [57] have carried out studies on reactive blending of polyamides to improve mechanical and thermal properties. They examined blends of PA with styrene–maleic anhydride copolymer (SMA). One aim of this work was to increase the heat distortion temperature (HDT) of PA. This was achieved with various additions levels of SMA. For example a 12 °C increase in HDT was obtained with the addition of 15% by weight of SMA to PA. Melt viscosity and elasticity of the PA were also increased on SMA addition. All of these effects were attributed to an anhydride-amine reaction occurring between the two materials.

Other similar reactions have been reported to produce compatible polyamide blends, for example PA reactive blends with polyesters. An amide–ester exchange occurs between these polymers, however the reaction is slow and in order to benefit from this reaction in terms of compatibility the reaction rate

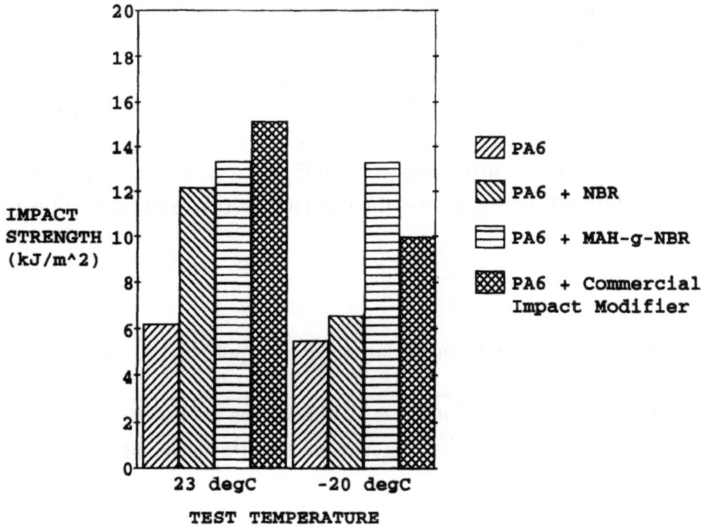

Figure 3.12 Izod impact strengths of nylon 6 blends containing impact modifiers, showing effectiveness of maleic anhydride-grafted NBR at low temperatures (Bonner *et al.* [56]).

must be increased. This increase in rate has been achieved using an organic phosphite catalyst [58].

3.5.3 Vinyl polymer reactive blends

In order to modify the mechanical properties and perhaps also to improve the solvent resistance of polystyrene, polyethylene can be blended with it. However, because of the incompatibility of these two polymers such a blend does not necessarily lead to these improved properties. By choosing a modified polystyrene and polyethylene which would react together, the improvement in properties can be achieved. Baker and Saleem [59] produced blends of acid modified LDPE and oxazoline modified PS in a melt mixer, giving materials with improved mechanical properties over unmodified polystyrene/polyethylene blends. For example, elongations to break were significantly higher for the reactive blend systems (see Table 3.3). The significant increase in elongation to break for the reactive blends was attributed to the formation of an *in situ* compatibiliser from the interaction between the two blend components. The reaction scheme is shown in Figure 3.13.

A similar reactive blending approach to that above has been investigated by Fowler and Baker [60] to rubber toughen polystyrene. Carboxylated nitrile rubber (CNBR) was blended with oxazoline modified polystyrene (OPS) in a melt mixer. Viscosity increases were observed for the reactive blend system with increased levels of oxazoline and carboxylic acid groups, indicating that a polymer–polymer interaction was occurring. In general, the impact strengths of the blends were found to be significantly higher than those for the non-reactive blends, supporting the view that an interaction was occurring similar to that shown in Figure 3.13.

From the viewpoint of examining the structural and rheological characteristics of reactive blends Hope *et al.* [61] have prepared and characterised a series of blends of ethylene vinyl alcohol (EVOH) and styrene maleic anhydride (SMA) copolymers containing up to 60% by weight EVOH. These were produced by extrusion using a co-rotating twin-screw extruder. The blends

Table 3.3. Elongation to break for polystyrene/polyethylene blends (after [59])

Proportion of modified PS and PE components in blend (wt/wt)	Elongation to break (%)	
	Acid modified PE	Unmodified PE
40/60	45	2
30/70	100	5
20/80	160	10
10/90	260	90
0/100	430	430

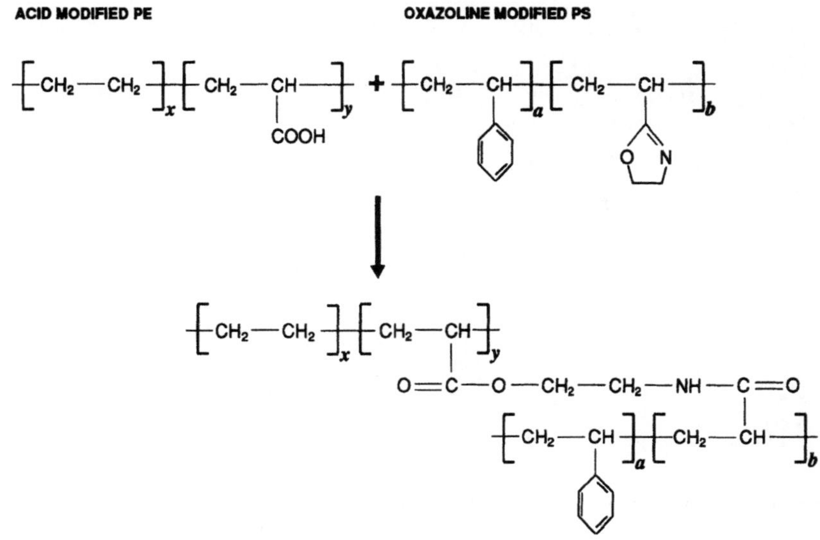

Figure 3.13 Proposed reaction between acid-modified PE and oxazoline-modified PS.

were characterised by Fourier transform infrared spectroscopy (FTIR), scanning electron microscopy, differential scanning calorimetry and capillary rheometry. In all cases a reaction was considered to have taken place as a carbonyl band was observed by FTIR for each blend and the viscosities of the blends were seen to increase in comparison to the individual blend components. The degree of reaction between the polymers, as inferred from the FTIR carbonyl peak, increased with EVOH content. At EVOH concentrations up to 50% by weight the EVOH existed as dispersed droplets in a SMA matrix, while the blend with 60% EVOH contained co-continuous phases. Examples of scanning electron micrographs are shown in Figure 3.14.

The viscous flow behaviour of the blends appeared to be dominated by the SMA matrix, but viscosity increased with EVOH content and with bonding level, and was always greater than that of the blend components (see Figure 3.15). The 60% EVOH blend also had a surprisingly low melt index, which may have been related to the co-continuous phase structure.

Datta *et al.* [62] have utilised reactive blending to compatibilise and toughen styrene–maleic anhydride copolymers by the addition of an amine functionalised ethylene–propylene polymer. It was proposed that a compatibiliser for this system was formed *in situ* via a reaction between the amine and the anhydride. The addition of up to 25 pph of the amine functional EP resulted in at least a ten-fold increase in notched Izod impact resistance of SMA copolymers at room temperature and at −40 °C. However, as might be expected, a decrease in modulus was observed with the addition of the rubbery EP component.

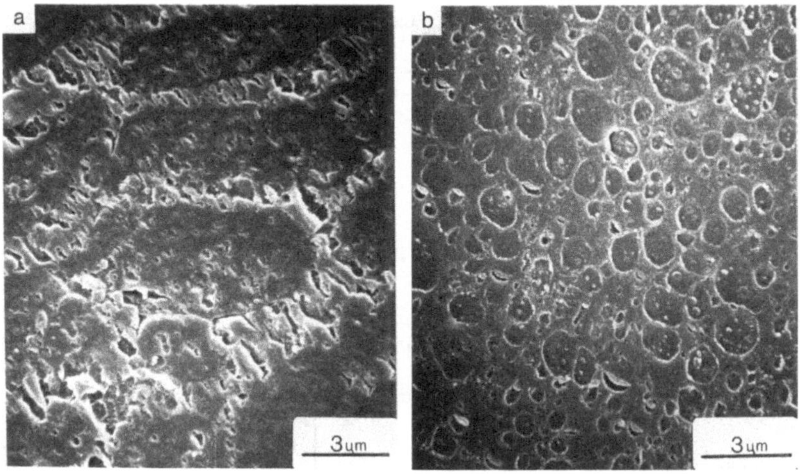

Figure 3.14 Scanning electron micrographs of EVOH/SMA blends: (a) 60/40 (b) 50/50 (after Hope *et al.* [61]).

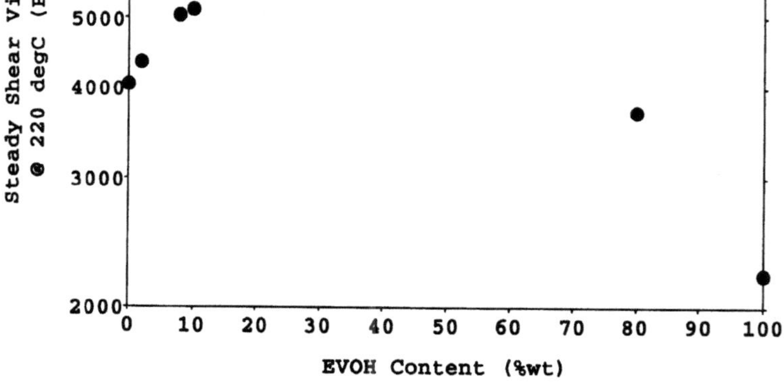

Figure 3.15 Viscosity of EVOH/SMA blends, measured by capillary rheometry at 220 °C (after Hope *et al.* [61]).

Unlike the other examples previously cited, where covalent bonding occurs during reactive blending, Simmons and Baker [63] have investigated a system where a polar interaction was considered to have taken place. A dimethylamino ethyl methacrylate grafted polyethylene was blended with a styrene–maleic anhydride copolymer. No significant viscosity increase was observed during blending and no new bands were detected in these blends by infrared spectroscopy, therefore it was concluded that no covalent bonding had occurred. However, scanning electron microscopy showed that blends containing the amine modified polyethylene were much more homogeneous than blends prepared with unmodified PE. This improved compatibility was attributed to strong polar interactions.

Other approaches to compatibilisation via a reactive processing route have been attempted. One approach was to blend incompatible polymers in the presence of a free radical initiator, such as a peroxide, or to irradiate the blend. The aim is for the initiator to attack the polymers creating polymer radicals which in turn may combine to form a copolymer compatibiliser *in situ*. For example, work has been carried out using peroxide as an additive to generate free radicals, by Cheung *et al.* [64], in PE/PP blends. However, this appeared to meet with limited success in that the addition of peroxide did not seem to have a significant compatibilising effect. However, earlier work by Rizzo *et al.* [65] found that irradiation did apparently give some compatibilisation in a similar blend system.

Another reactive approach has been to make use of electronic interactions to promote compatibility. Molnar and Eisenberg [7] studied this with blends of sulphonated polystyrene and polyamide, and with blends of lithium neutralised sulphonated PS and PA. In both cases compatibility was enhanced compared to a PS/PA blend, as shown by a decrease in the heterogeneity of the blends and in the glass transitions of the blend components merging together.

3.6 Future trends

Compatibilisation of blend components is a major consideration when designing blends and is often the primary criterion for commercial success. Hence much of the compatibilisation technology which exists is proprietary information. This situation is unlikely to change in today's increasingly competitive marketplace.

The major future challenges and developments for compatibilisation technology lie in three main areas of blending: engineering polymer blends, superior performance commodity polymers and polymer recycling. In the first two areas the primary objective is to produce a material where some specific property is enhanced. This blending option is very often a significantly more cost-effective route to producing an improved material than developing a completely new polymer. Development costs for a blend can be relatively low and capital investment minimal, particularly if compounding equipment already exists.

Many of the developments in engineering polymer blends are aimed at applications where very high performance is required, such as aerospace products. For example an engineering polymer blend which appeared in recent patent literature was a polyetherimide/copolyester liquid crystal polymer combination, where a material with high heat distortion temperature and flexural modulus was produced. Although the mechanism was not disclosed it seems likely that compatibilisation in this system is achieved via a similar reactive route to that used in polycarbonate/polyester blends. This compatibilisation route seems an attractive one which may merit development for other systems in the future.

In terms of improving the properties of commodity polymers some blends have already been discussed in this chapter. The properties to be improved include mechanical properties, like toughness and stiffness, thermal properties such as melting and softening points, permeation performance and chemical resistance. Applications for improved performance commodity polymers include automotive parts, business machines, hand tools, consumer products and household appliances. Specific examples of such blends include polyamide/polyethylene or polypropylene combinations where compatibilisation is achieved by addition of a third component, a functional polyolefin. These blends can exhibit improved thermal properties over unmodified polyolefins and, as discussed earlier, improved permeation resistance to solvents. It is likely that development activity on commodity polymer blend systems, and hence suitable compatibilisation routes, will increase, particularly for polyolefin based materials in automotive applications, where polyolefins are favoured from the point of view of recyclability [19].

The area of polymer recycling itself is a very important issue for the industry today. There are environmental pressures on industry to recycle polymers, particularly those used in packaging applications, which are often discarded as waste. Most post-consumer waste streams are mixtures of many different polymers and if recycling of this waste is carried out then very often the production of a multicomponent blend will result, which is technically and economically more favourable than separation of the waste into its constituent polymers. Production of such blends will, however, provide significant challenges in terms of compatibilisation, as a result of the potential complexity of the mixture, which may contain polyethylene, polypropylene, PVC, polystyrene, polyamide and other polymers, such as barrier polymers. Although many of the compatibilisation routes discussed earlier could be utilised to some extent, more versatile compatibilisers may be required. Development of these compatibilisers is taking place and some are already commercially available. One such material is Bennett, which is produced by Bennett BV and marketed as a 'universal compatibiliser'. Although the structure of Bennett has not been disclosed it appears to be a thermoplastic polymer containing a range of different functional groups. This facilitates reaction with a number of polymers to form block copolymers *in situ*, which in turn have a compatibilising effect in polymer mixtures [24].

As the number of blends grows, as is predicted, then the need for specific compatibilisation routes will continue to be a much researched field. In addition to producing universal compatibilisers, like Bennett, specific compatibilisers will certainly continue to be developed. One approach of growing interest is molecular modelling, in which molecular response is modelled with respect to the molecule's immediate local environment. This may in future allow tailoring of molecular structure to exploit specific interactions, which in turn may allow the development of 'designer' compatibilisers.

References

1. Krause, S. (1978) in *Polymer Blends*, Vol. 1, Paul, D.R. and Newman, S. eds., Academic Press, London.
2. Subramanian, P.M. and Mehra, V. (1987) *Polym. Eng. Sci.* **27**, 633.
3. Lester, A.J. and Hope, P.S. (1987) *European Symposium on Polymer Blends*, Strasbourg, May 25–27.
4. Barentsen, W.M., Heikens, D. and Piet, P. (1974) *Polymer* **15**, 119.
5. Fayt, R., Jerome, R. and Teyssie, Ph. (1986a) *J. Polym. Sci., Polym. Lett. Edn.* **24**, 25.
6. Fayt, R., Jerome, R. and Teyssie, Ph. (1986b) *Makromol. Chem.* **187**, 837.
7. Molnar, A. and Eisenberg, A. (1991) *Polymer Communications* **32**(12), 370.
8. Fayt, R., Jerome, R. and Teyssie, Ph. (1981a) *J. Polym. Sci., Polym. Lett. Edn.* **19**, 79.
9. Fayt, R., Jerome, R. and Teyssie, Ph. (1981b) *J. Polym. Sci., Polym. Phys. Edn.* **19**, 1269.
10. Fayt, R., Jerome, R. and Teyssie, Ph. (1982) *J. Polym. Sci., Polym. Phys. Edn.* **20**, 2209.
11. Fayt, R., Jerome, R. and Teyssie, Ph. (1989) *J. Polym. Sci., B. Polym. Phys.* **27**, 775.
12. Paul, D.R. (1978) in *Polymer Blends*, Vol. 2, Paul, D.R. and Newman, S. eds., Academic Press, London.
13. Gaylord, N.G. (1975) in *Copolymers, Polyblends and Composites*, Platzer, N.A.J., ed., Amer. Chem. Soc., Washington DC.
14. Nolley, E., Barlow, J.W. and Paul, D.R. (1980) *Polym. Eng. Sci.* **20**, 364.
15. Ho, W.-J. and Salovey, R. (1981) *Polym. Eng. Sci.* **21**, 839.
16. Bartlett. D.W., Barlow, J.W. and Paul, D.R. (1982) *J. Appl. Polym. Sci.* **27**, 2351.
17. Heikens, D., Hoen, N., Barentsen, W., Piet, P. and Ladan, H. (1978) *J. Polym. Sci., Polym. Symp.* **62**, 309.
18. Paul, D.R., Locke, C.E. and Vinson, C.E. (1973) *Polym. Eng. Sci.* **13**, 202.
19. Higuchi, S., Akiba, T., Ishidate, Y. and Miura, T. (1991) *Proc. Compalloy Europe '91.*
20. Barentsen, W.M. and Heikens, D. (1973) *Polymer* **14**, 579.
21. Barentsen, W.M., Heydenrijk, P.J., Heikens, D. and Piet, P. (1978) *Brit. Polym. J.* **10**, 17.
22. Locke, C.E. and Paul, D.R. (1973a) *J. Appl. Polym. Sci.* **17**, 2791.
23. Locke, C.E. and Paul, D.R. (1973b) *Polym. Eng. Sci.* **13**, 308.
24. Van der Groep, L.A. (1991) *Proc. Compalloy Europe '91.*
25. Miles, A.F., Bonner, J.G. and Hope, P.S. (1992) (in prep.).
26. Brahimi, B., Ait-Kadi, A., Ajji, A., Jerome, R. and Fayt, R (1991) *J. Rheol.* **35**, 1069.
27. Dumoulin, M.M., Farha, C. and Utracki, L.A. (1984) *Polym. Eng. Sci.* **24**, 1319
28. Yang, D., Zhang, B., Yang. Y., Fang, Z., Sun, G. and Feng, Z. (1984) *Polym. Eng. Sci.* **24** (8), 612.
29. Lohse, D.J., Datta, S. and Kresge, E.N. (1991) *Macromolecules* **24**, 561.
30. Ide, F. and Hasegawa, A. (1974) *J. Appl. Polym. Sci.* **18**, 963.
31. Cimmino, S., D'Orazio, L., Greco, R., Maglio, G., Malinconico, M., Mancarella, C., Martuscelli, E., Palumbo, R. and Ragosta, G. (1984) *Polym. Eng. Sci.* **24**, 48.
32. Chen, C.C., Fontan, E., Min, K. and White, J.L. (1988) *Polym. Eng. Sci.* **28** (2), 69.
33. Kim, B.K., Park, S.Y. and Park, S.J. (1991) *Eur. Polym. J.* **27**, 349.
34. Meyer, R.V. and Dhein, R. (1977) *Chem. Abstr.* **87**, 185771.

35. Wingler, F. and Liebig, L. (1978) *Chem. Abstr.* **88**, 8001.
36. Lindberg, K.A.H., Johansson, M. and Bertilsson, H.E. (1990) *Plast. Rub. Proc. Appl.* **14**, 195.
37. Angola, J.C., Fujita, Y., Sakai, T. and Inoue, T. (1988) *J. Polymer Sci., Polym. Phys. Edn.* **26**, 807.
38. Schramm, J.N. and Blanchard, R.R. (1970) *SPE RETEC*, New Jersey, October.
39. Wollrab, F., Declerck, F., Dumoulin, J., Obsomer, M. and Georlette, P. (1972) *Am. Chem. Soc., Polym. Div. Prepr.* **13** (1), 499.
40. Lindsay, C.R., Paul, D.R. and Barlow, J.W. (1981) *J. Appl. Polym. Sci.* **26**, 1.
41. Traugott, T.D., Barlow, J.W. and Paul, D.R. (1983) *J. Appl. Polym. Sci.* **28**, 2947.
42. Endo, S., Min, K., White, J.L. and Kyu, T. (1986) *Polym. Eng. Sci.* **26**, 45.
43. Kato, K. (1967) *Polym. Eng. Sci.* **8**, 38.
44. Fiese, K. (1971) *Plaste u. Kautschuk* **18**, 881.
45. Noolandi, J. and Hong, K.M. (1982) *Macromolecules* **15**, 482.
46. Marie, P., Selb, J., Rameau, A., Duplessix, R. and Gallot, Y. (1985) in *Polymer Blends and Mixtures*, Walsh, D.J., Higgins, J.S. and Maconnachie, A., eds., Martinus Nijhoff, Dordrecht.
47. Reiss, G., Kohler, J., Tournut, C. and Banderet, A. (1967) *Makromol. Chem.* **101**, 58.
48. Kohler, J., Reiss, G. and Banderet, A. (1968) *Eur. Polym. J.* **4**, 173.
49. Inoue, T., Soen, T., Hashimoto, T. and Kawai, H. (1970) *Macromolecules* **3**, 87.
50. Bartlett, D.W., Paul, D.R. and Barlow, J.W. (1981) *Mod. Plast.* **58**, 60.
51. Del Guidice, L., Cohen, R.E., Attalla, G. and Bertinotti, F. (1985) *J. Appl. Polym. Sci.* **30**, 4305.
52. Robeson, L.M., Matzner, M., Fetters, L.J. and McGrath, J.E. (1974) in *Recent Advances in Polymer Blends, Grafts and Blocks*, Sperling, L.H. ed., Plenum, New York.
53. Devaux, J., Goddard, P. and Mercier, J.P. (1982) *J. Polym. Sci., Polym. Phys. Edn.* **20**, 1901.
54. Robeson, L.M. (1985) *J. Appl. Polym. Sci.* **30**, 4081.
55. Sjoerdsma, S.D. (1989) *Polymer Communications* **30**, 106.
56. Bonner, J.G., Hope, P.S. and Jackson, A.S. (1992) (in prep.).
57. Kim, B.K. and Park, S.J. (1991) *J. Appl. Polym. Sci.* **43**, 357.
58. Aharoni, S.M. (1985) in *Integration of Fundamental Polymer Science and Technology*, Kleitjens, L.A. and Lemstra, P.J., eds., Elsevier, London and New York.
59. Baker, W.E. and Saleem, M. (1987) *Polym. Eng. Sci.* **27** (20), 1634.
60. Fowler, M.W. and Baker, W.E. (1988) *Polym. Eng. Sci.* **28** (21), 1427.
61. Hope, P.S., Bonner, J.G. and Curry, J. (1992) *IUPAC International Symposium on Macromolecules*, Prague, 7–10 July 1992.
62. Datta, S., Dharmarajan, N., Ver Strate, G. and Kaufman, L.G. (1991) New concepts in impact modification of styrene maleic anhydride resins. *Proc. Compalloy Europe '91.*
63. Simmons, A. and Baker, W.E. (1990) *Polymer Communications* **31**, 20.
64. Cheung, P., Suwanda, D. and Balke, S.T. (1990) *Polym. Eng. Sci.* **30** (17), 1063.
65. Rizzo, G., Spadaro, G., Acierno, D. and Calderaro, E. (1983) *Radiat. Phys. Chem.* **21**, 349.

4 Rheology of polymer blends

J. LYNGAAE-JØRGENSEN

4.1 Introduction

4.1.1 *Definitions* [1]

Polymer blend (*PB*). A mixture of at least two polymers or copolymers.

Miscible PB. PB homogeneous down to molecular level: $\Delta G_{\mathrm{m}} \cong \Delta H_{\mathrm{m}} \leqslant 0$.

Immiscible PB. $\Delta G_{\mathrm{m}} \cong \Delta H_{\mathrm{m}} > 0$.

Compatible PB (Utilitarian term). Homogeneous to the eye, commercially attractive PB.

Polymer alloy (*PA*). Immiscible PB having a modified interface and/or morphology.

Polymer materials may be grouped in different ways. An overview of the most common two-phase systems consisting of polymers is shown in Figure 4.1. This chapter deals mainly with melt flow properties of physical mixtures of different polymers. The rheology of block copolymers is reviewed in [2]. Rheology is the science of the deformation and flow of materials. For polymers there may be at least three distinct objectives of rheological work:

(a) To develop fundamental understanding of material behaviour in order to develop rheological equations of state.
(b) To use the simplest possible laboratory investigations in order to find relationships with processing practice in order to govern and predict processing properties of polymer materials.
(c) To develop relationships between measures of polymer structure and properties, in this case rheological properties. If such relationships have been established, rheological data can be used for characterisation of, say, average molecular weight, determination of branching in polymers, etc.

The establishment of valid rheological equations is an active research field. A very large body of possible rheological equations has been established

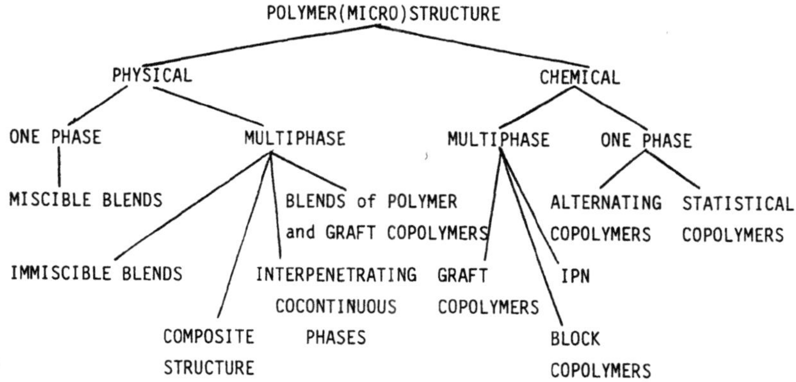

Figure 4.1 Microstructure of polymer blends and alloys.

[3,4]. The behaviour of a polymer during flow is not easily predicted. A general rheological equation must be formulated, the right boundary conditions for a specific case established, the problem defined and then solved. This process is not straightforward because polymer compounds are normally much more complex than pure homopolymers and the limiting conditions for flow, e.g. melt fracture, the existing rheological equations, etc., cannot be predicted accurately.

In many commercially available CAE programs, e.g. 'Mould Flow', very simple rheological equations often based on steady-state flow measurements in simple flow are used.

4.2 Experimental measuring data obtained in simple flow fields

As mentioned above, the evaluation of flow behaviour may be confined to a few well-defined simple flow cases in order to make an attempt to interpret more easily the anomalies of thermorheologically complex systems [5]. The definitions of the applied rheological functions are outlined briefly in the following sections. Comprehensive treatments may be found in many textbooks on rheology [3–5].

4.2.1 The steady shearing flow field

Consider a flow geometry consisting of two infinitely long, parallel planes forming a narrow gap filled with a fluid. The plates are separated by distance H, which is very small compared with the width, W, of the plane (i.e. $H \ll W$), as shown schematically in Figure 4.2. The velocity field of the laminar flow at steady-state between the two planes is given by

$$v_x = f(y), \quad v_z = v_y = 0$$

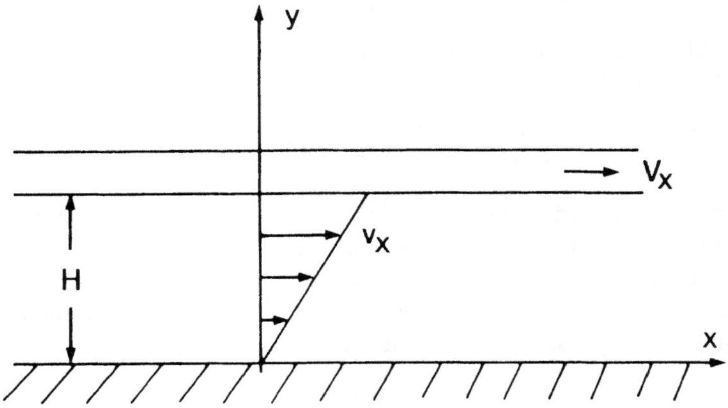

Figure 4.2 Schematic diagram of simple shearing flow.

Referring to Figure 4.2, the lower plane is stationary (i.e. $v_x(0)=0$), and the upper plane moves in the x direction at a constant speed, v_x, i.e. $v_x(H)=V_x$. The velocity gradient, dv_x/dy is then constant, i.e.

$$\dot{\gamma} = dv_x/dy = \text{constant} \tag{4.1}$$

where $\dot{\gamma}$ is the shear rate.

In steady simple shearing flow the stress tensor may be written

$$\sigma = \begin{Bmatrix} \sigma_{xx} & \sigma_{xy} & 0 \\ \sigma_{yx} & \sigma_{yy} & 0 \\ 0 & 0 & \sigma_{zz} \end{Bmatrix} \tag{4.2}$$

where the extra stress tensor $\sigma = S + P \cdot \delta$ (S is the total stress tensor, P is the hydrostatic pressure and δ is the unit tensor). The subscript x denotes the direction of flow in Figure 4.2, y denotes the direction perpendicular to the flow and z denotes the neutral direction. $\sigma_{xy} = \sigma_{yx}$ is the shear stress and σ_{xx}, σ_{yy} and σ_{zz} are normal stresses.

Three material functions are sufficient to describe the stress–shear rate dependence in simple shear flow. These are normally defined by

$$\sigma_{yx} = \eta(\dot{\gamma})\dot{\gamma}, \quad \sigma_{xx} - \sigma_{yy} = \psi_1(\dot{\gamma})\dot{\gamma}^2$$

and

$$\sigma_{yy} - \sigma_{zz} = \psi_2(\dot{\gamma})\dot{\gamma}^2$$

where $\eta(\dot{\gamma})$ is the viscosity function and $\psi_1(\dot{\gamma})$ and $\psi_2(\dot{\gamma})$ are the first and second normal stress coefficients, respectively. The experimental techniques most used for the determination of the rheological functions in simple shear involve the use of capillary viscometers (Poiseuille flow), coaxial cylinder viscometers (Couette flow) and cone and plate viscometers [6, 7].

4.2.2 Dynamic data (oscillatory shear flow)

The displacement gradient dx/dy (Figure 4.2) is known as the shear strain and given the symbol γ

$$\gamma = \frac{dx}{dy} = \text{shear strain (dimensionless)}$$

The time rate of change of shear strain $\dot{\gamma}$ (the dot is Newton's notation for the time derivative) is the shear rate given in eqn (4.1).

If a linear viscoelastic fluid is subjected to a sinusoidally varying shear strain, γ in Figure 4.2, the measured stress, σ, will also vary sinusoidally but normally out of phase with the strain, i.e.

$$\gamma = \gamma_0 \cos \omega t \tag{4.3}$$

$$\sigma = \sigma_0 \cos (\omega t + \delta) \tag{4.4}$$

where γ_0 and σ_0 are the maximum values of strain and stress, respectively, ω is the circular frequency, t is time and δ is the phase angle.

The analysis of dynamic data is most easily performed by introducing complex quantities. A complex shear strain and shear stress may be defined as follows:

$$\gamma^* = \gamma_0(\cos \omega t + i \sin \omega t) = \gamma_0 \exp(i\omega t) = \gamma' + i\gamma'' \tag{4.5}$$

and

$$\sigma^* = \sigma_0(\cos(\omega t + \delta) + i(\sin \omega t + \delta)) = \sigma_0 ep(i(\omega t + \delta))$$

where

$$|\sigma^*| = \sigma_0 \text{ and } \sigma = \text{Re}\, \sigma^* \tag{4.6}$$

Using the definitions given in eqns (4.5) and (4.6), a complex modulus,

$$G^* \equiv \frac{\sigma^*}{\gamma^*} = G' + iG'' \tag{4.7}$$

and a complex viscosity

$$\eta^* = \frac{\sigma^*}{\dot{\gamma}^*} = \frac{\sigma^*}{i\omega\gamma^*} = \eta' - i\eta'' \tag{4.8}$$

both of rheological significance, may be defined.

Thus

$$G^* = \frac{\sigma_0 \exp(i\omega t + \delta)}{\gamma_0 \exp(i\omega t)} = \frac{\sigma_0}{\gamma_0}(\cos\delta + i\sin\delta) \text{ or } G' = \frac{\sigma_0}{\gamma_0}\cos\delta$$

and

$$G'' = \frac{\sigma_0}{\gamma_0}\sin\delta$$

and

$$\eta^* = \frac{\sigma^*}{i\omega\gamma^*} = \frac{i}{i}\left[\frac{G'}{i\omega} + \frac{iG''}{i\omega}\right] = \frac{G''}{\omega} - \frac{iG'}{\omega}$$

or

$$\eta' = \frac{G''}{\omega} \text{ and } \eta'' = \frac{G}{\omega}$$

where $\eta'(\omega)$ is the dynamic viscosity and $G'(\omega)$ is the storage modulus. G' and η'' can be shown to be associated with energy storage and release. $G''(\omega)$ is called the loss modulus. G'' and η' are associated with the dissipation of energy as heat.

A large number of instruments can be used to measure these rheological functions [6, 7].

4.2.3 *Uniaxial elongational flow*

A flow field of great practical importance is the uniaxially extensional flow that may be found in such polymer fabrication processes as fibre spinning and processing which involve converging entry flows [8]. For uniaxial stretching (e.g. fibre spinning) the velocity field is given as

$$v_x = f(x), \quad \frac{\partial v_x}{\partial x} + \frac{\partial v_y}{\partial y} + \frac{\partial v_z}{\partial z} = 0 \tag{4.9}$$

Equation (4.9) assumes a flat velocity profile in the directions perpendicular to the flow direction. For such a flow field, the rate-of-strain tensor $\dot{\gamma}$ is

$$\dot{\gamma} = \begin{Bmatrix} \dot{\gamma}_E & 0 & 0 \\ 0 & -\dot{\gamma}_E/2 & 0 \\ 0 & 0 & -\dot{\gamma}_E/2 \end{Bmatrix}$$

in which $\dot{\gamma}_E$ is the elongation rate, defined as

$$\dot{\gamma}_E = dv_x/dx \tag{4.10}$$

For uniaxial elongation flow, the elongational viscosity η_E may be defined by the ratio of tensile stress S_{xx} and rate of elongation (or elongation rate) $\dot{\gamma}_E$,

$$\eta_E = S_{xx}/\dot{\gamma}_E \tag{4.11}$$

If the surfaces transverse to the direction of principal elongation (i.e. the direction of stretching) are unconstrained, then

$$S_{yy} = S_{zz} = 0 \tag{4.12}$$

and eqn (4.11) may be rewritten as

$$\eta_E = (\sigma_{xx} - \sigma_{yy})/\dot{\gamma}_E$$

Over the last few years a large number of measuring instruments have been developed which measure η_E [6, 7].

For polymer blends (PB) the question of valid rheological equations is simply not answerable at present. What we can do is evaluate the flow behaviour of polymer blends in simple well-defined geometric flows, e.g. in simple shear flow and in simple elongation. The processing properties may be evaluated from these data by simple techniques such as dimensional analysis and similitude studies.

In this chapter flow behaviour of PBs in the melt state will be treated with emphasis on structuring, i.e. governing of two-phase structure by control of the flow prehistory.

4.3 Miscibility and flow behaviour of polymer blends

For polymer blends a distinction must be made between miscible blends and immiscible blends. A necessary condition for miscibility is

$$\Delta G_{mix} = \Delta H_m - T\Delta S_m \leqslant 0$$

A sufficient condition for miscibility in all proportions is then that

$$\delta\Delta^2 G_m/(\delta\varphi_2)^2 > 0$$

When the molecular weight of the polymers in the blend increases ΔS_m becomes small or zero ($\bar{x}_n \rightarrow \infty \rightarrow \Delta S_m \rightarrow 0$). Miscible polymer blends show flow behaviour which is equivalent with homopolymer flow behaviour. The homologous polymer mixtures are good model systems for miscible polymer blends. Over the years many of these have been studied, providing an invaluable source of information. However, phase behaviour and phase transitions may be induced by flow [1, 9].

There are then three cases: completely miscible systems during flow, systems with phase transitions during flow, and completely immiscible systems.

4.3.1 *Miscible polymer blends*

For completely miscible systems the 'equivalent' flow behaviour for polymer blends and for homopolymers may be expected and predicted. Of course there are many interesting questions, for example, are the systems homogeneous at all magnitudes of volume and influence on entanglement density. However, it seems correct that from a macrorheological point of view characteristic flow properties for a given composition show rheological behaviour equivalent with that found for homopolymers and that measurements of rheological behaviour will normally be functions of volume fraction alone [1].

4.3.2 *Systems exhibiting phase transitions during flow*

The materials showing flow induced structure/phase changes are an interesting group. However, these phenomena are marginal for polymer blends in the sense that only systems which are nearly miscible will show such transitions [9].

Phase transitions in polymer blends and block copolymers. Theoretically, it is predicted that a homogeneous state may be formed during simple flow above a critical shear stress or shear rate which depends primarily on the miscibility of the two polymers. In our laboratory light scattering measurements on a (nearly miscible) model system consisting of mixtures of SAN and PMMA were applied during flow in order to evaluate the prediction that a phase transition to a homogeneous state could be provoked by application of a critical shear stress. At conditions within the spinodal range of the phase diagram (corresponding to approx. 20 °C temperature difference) it was shown that above the stress $\sigma_c \cong 80$ kPa the total scattering is approximately zero and the anisotropic scattering pattern disappears [10]. When the shearing is stopped the scattering pattern is independent of scattering angle and shows maxima in a plot of scattering intensity as a function of scattering angle [11]. The structure formed is an interpenetrating co-continuous structure. The melt is optically clear above σ_c and milky below.

Thus it has been documented that a phase transition to homogeneous state may be provoked by shearing. All literature data reported to date show that homogenisation may take place at sufficiently high shear stress or shear stress [10–22].

Since the applicable stress is limited by melt fracture phenomena, this way of producing blends with dual continuity is only possible for nearly miscible blends. This case is thus mainly interesting for mixtures of tailor-made copolymers.

A summary of the area and its possibilities might be as follows: even though hundreds of miscible blends have been reported, in general a blend of two different high molecular polymers will be immiscible and thus form a two-phase system.

For two-phase blends the general conclusion from a very large number of investigations is that the properties of an immiscible blend are a function of volume fraction and selected measures characterising the two-phase structure. This is in contrast with the behaviour of miscible blends.

Neither the time–temperature superposition of shear viscosity nor of extrudate swell is observed for non-compatibilised immiscible blends. The lack of superposition is a rule not an exception. There are several reasons for this. The temperature dependence of shear viscosity varies from one polymer to the next, which leads to different viscosity ratios, λ, at different temperatures. The difference λ also results in variation of the extrudate morphology, i.e. different drop deformation, different degree of shear segregation, etc., which by reciprocity changes the rheological response of the system.

When dealing with rheology of two phase systems it is then normally necessary to include a microrheological analysis or point of view in the treatment of flow data.

4.4 Flow behaviour of immiscible polymer blends

Suspensions, emulsions or block copolymers have served as models for immiscible polymer blends. Since the two-phase structure influences the properties of immiscible PB, it is expected that these materials will show complex and time dependent behaviour. In these cases it is necessary to make the distinction between steady-state and transient properties, especially if comparisons between different measurements are made.

4.4.1 Macrorheological data

Polymer blends can be treated as a continuous homogeneous phase system and in simple shear flow, for example, an apparant shear stress and shear rate can be measured. Such data are shown in Figure 4.3.

The steady-state data look superficially equivalent with data for homopolymers. However, time–temperature superpositions are not generally followed and viscosity composition data (e.g. showing both minima and maxima), mixing rules, time dependencies as well as die swell behaviour can be very complex indeed. Rationalisation of this behaviour as well as two-phase structure development involves a microrheological understanding.

4.4.2 Microrheological approach

To start with, a short reference to the fundamental factors influencing the properties of polymers is given. In Figure 4.4 measures of fundamental molecular characteristics and microstructural characteristics are reviewed.

For polymer blends characterisation involves the measurement of volume fraction, and the determination of the two-phase structure and interfacial characteristics. The question of the necessary (and sufficient) number of measures of structure which sufficiently characterise a two-phase structure in order to predict a given property are far from trivial and certainly not solved in general terms.

The microrheology of two-phase systems has been approached from different angles. One method involves Newtonian drops in Newtonian media. This approach starts from infinite dilution by equivalence with emulsion and/or suspension behaviour [1]. In general we want to consider blends of two incompatible uncrosslinked molten polymers, forming a dispersion. The rheological behaviour of such blends depends on:

(a) *Rheological behaviour of each phase:*
Polymer forming the matrix: η_2, $N_{1,2}$, $G_2^*(\omega)$, etc.
Polymer forming the dispersed inclusions: η_1, $N_{1,1}$, $G_1^*(\omega)$, etc.

(b) *Morphology:*
The necessary, and hopefully sufficient, number of independent, discriminating valid quantitative structural data which enable the pre-

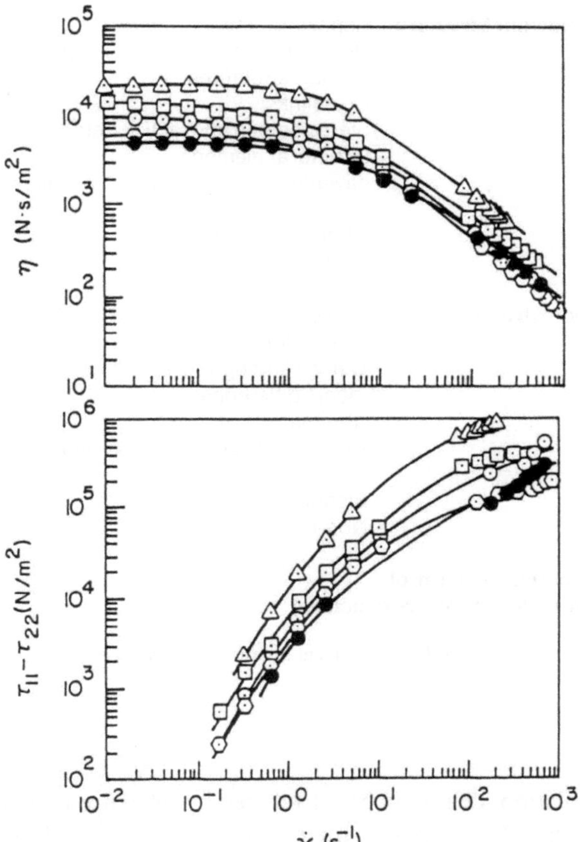

Figure 4.3 Viscosity and first normal stress difference versus shear rate for blends of polymethylmethacrylate (PMMA) and polystyrene (PS) at 200 °C [41]: ●, PS; △, PMMA; ⊡, PS/PMMA = 27/75 (by wt.); ⊙, PS/PMMA = 50/50 (by wt.); ⊙, PS/PMMA = 75/25 (reprinted with permission of Academic Press, Inc.).

diction of the properties of a blend which is to be simulated, must be determined. This basic problem is not satisfactorily solved at present.

A reasonable start is to distinguish between systems with discrete domains of one phase in a matrix phase and systems with dual phase continuity. We define the so-called degree of continuity of each phase: φ_A = the degree of continuity for phase A (or φ_B for phase B), i.e. the volume fraction of phase A (or phase B, respectively) which is part of an infinite (percolation) structure [25]. The main reason for these definitions is that general property structure relationships developed in the so-called percolation theory seem to be applicable for polymer blends (e.g. for mechanical moduli) [25, 26].

1. Molecular Structure – Repetition unit
- Distribution of molecular lengths
- Branch structures
- Crosslinks
- Stereoisomerism (tacticity etc.)
- Structural elements deviating from the repetition unit
  - Comonomers
  - End to end structures
  - End groups etc.

2. Microstructure
- Defects
- Orientation
- Compounding ingredients

"Local" differences in the material matrix:
- e.g. Distinct phases – two phase systems
- Crystallinity
- Blends
- Non-mixable agents

3. Geometrical form of the material (macrostructure)

Figure 4.4 Fundamental structural variables.

Percolation ideas may be introduced by reference to a now classical experiment. Two types of materials are mixed in a beaker and the electrical conductance through the mixture between the bottom and the top of the beaker is measured. The mixture consists of small beads of an electrical isolator, e.g. glass beads, and a conducting material, e.g. silver beads. As long as the metal beads form isolated structures no current is measured. At a critical volume fraction, φ_c, a continuous structure of metal contacts percolates throughout the entire material, even when the total volume goes to infinity, and an electrical current is produced through the mixture. Roughly the same ideas have been applied to polymer blends [25, 26]. It has been shown that the percolation threshold volume, φ_c, depends heavily on the form of the discrete particles it is formed from but is close to the theoretical value ($\varphi_c = 0.156$) for spherical domains.

A satisfactory characterisation of discrete systems of simple form, such as spheres or ellipsoids, includes estimates of volume fraction of discrete phase and distribution of radii or of small and large semi-axis and their orientational distribution. Even for more complex but still discrete domains satisfactory characterisation is relatively straightforward.

However, a valid characterisation of the three-dimensional structures found in blends with dual continuity structures represents both theoretical and practical measuring problems. One approach is to characterise the structure by fitting 'model parameters' which generate three-dimensional structures with, e.g. the same total interface area, characteristic dimensions and orientation distribution.

The optimal choice of structural characterising measures depends on the properties to be predicted. The problem is non-trivial but it is certain that the commonly used qualitative visual evaluation of e.g. electron microscopy pictures is often misleading. Quantitative structure evaluation by picture analysis, for example, is a necessity.

(c) *Interfacial properties:*
Interfacial tension v (depending on amount of compatibiliser at the interface) and thickness of interfacial layers.

4.5 Drop break-up in two-phase flow: Newtonian drop in Newtonian medium

In simple flow fields of discrete liquid drops, stresses arising in the continuous phase tend to deform and orient a drop [27]. The force responsible for the deformation is due to the difference, ΔP, between the pressures inside and outside the drop. For Couette flow, the pressure difference is expressed as [28]

$$\Delta P = 4\dot{\gamma}\eta_2 \left[\frac{19\lambda + 16}{16\lambda + 16}\right] \sin 2\phi,$$

where $\dot{\gamma}$ is the rate of shear; η_2 is the viscosity of the matrix phase; $\lambda = \eta_1/\eta_2$ is the ratio of the viscosities of the melt of the discrete phase η_1 and the medium η_2. ϕ is the angle determining the orientation of the drop in the direction of the flow.

At low shear rates the forces trying to disrupt a drop are balanced by the forces keeping it together and the drop assumes the shape of an ellipsoid. At high shear rates the drop deforms into a liquid cylinder and may break up. The restoring force exerted via the interphase tension, ΔP_v, may be determined by the Laplace equation:

$$\Delta P_v = \frac{2v}{r},$$

where v is the interphase tension and r is the radius of an undeformed drop. Note that

$$\Delta P_v = v\left[\frac{1}{r_1} + \frac{1}{r_2}\right]$$

for an ellipsoidal drop with small and large radius of curvature r_1 and r_2, respectively.

Under equilibrium conditions, the deformation D of the drop is given by

$$D = (L - B)/(L + B) = \frac{19\lambda + 16}{16\lambda + 16} \cdot \kappa$$

where κ is the capillarity number, and L and B are the major and minor axes of the deformed drop, respectively.

Thus, according to Taylor's theory, the deformation of a Newtonian drop in a Newtonian matrix, depends on two parameters: the viscosity ratio $\lambda = \eta_1/\eta_2$ of the drop and the medium and the capillarity number κ, i.e. the ratio of the product of the local shear stress and the drop radius to the interphase tension:

$$\kappa = \eta_2 \dot{\gamma} d / v$$

where $d = 2r$.

Taylor [27] supposed that the drop fractures when the surface tension, which seeks to keep the drop in spherical form, can no longer be balanced by viscous forces which attempt to break the drop. Therefore, the parameters λ and κ which control the deformation must also control the critical conditions for fracture. According to Taylor, the deformation D_c of the drop during fracture (critical value of deformation) is 0.5.

A large number of theoretical and experimental examinations have been carried out on the fracture of a Newtonian liquid cylinder in a steady uniaxial or shear flow, under transient conditions, and in a resting Newtonian fluid after the cessation of flow [28–40].

Rumscheidt and Mason [37] describe four types of drop fracture of Newtonian fluids according to the value of λ. For $\lambda < 0.5$ the drop becomes S-shaped and tiny drops break away from its ends at a critical value of $\dot{\gamma}$. As the viscosity of the drop increases ($\lambda > 0.2$), the deformed drop breaks through the centre with the formation of two large drops and three satellite drops. For $\lambda < 2$ the drop extends into a long cylinder and eventually falls into many small drops. For $\lambda > 2$ the drop becomes deformed and is oriented along the flow without fracture even at $\dot{\gamma}$ up to $40\,s^{-1}$.

The findings of most investigations [41] may be generalised in a plot of critical capillarity number against viscosity ratio as shown in Figure 4.5 for simple shear flow and uniaxial extensional flow, respectively. Below the curves deforming drops are stable and above they are unstable. The exact curves vary somewhat from author to author but the general picture is the same. An important generalisation is that drops cannot be fractured in simple shear flow if the viscosity ratio is larger than 3–4 [33, 39, 42] but may be fractured in extensional flow at all values of viscosity ratio.

The stability of liquid threads surrounded by an immiscible matrix as treated by Tomotika [29] may be given as follows. He assumed the thread to exhibit a small axisymmetric sinusoidal distortion, as is shown in Figure 4.6:

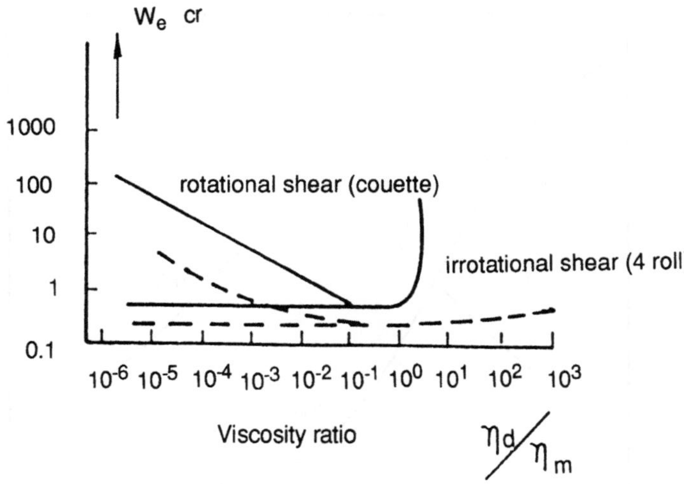

Figure 4.5 Capillarity number κ (Weber number) at burst versus viscosity ratio λ. — simple shear, --- plane hyperbolic (reprinted with permission of J. Elmendorph [50]).

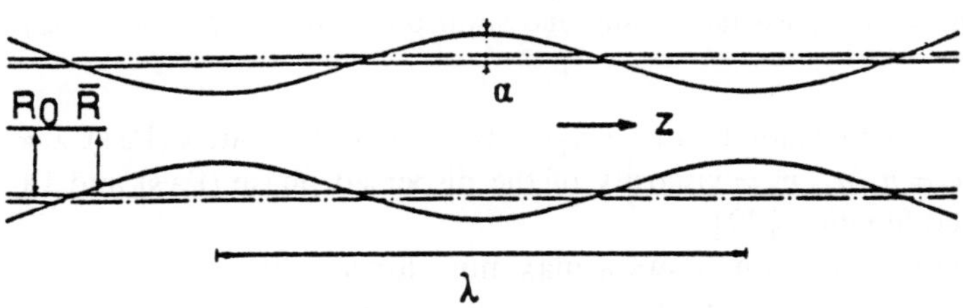

Figure 4.6 Sinusoidal Rayleigh distortion on a liquid cylinder (reprinted with permission of J. Elmendorph [50]).

$$R(z) = \bar{R} + \alpha \sin\left[\frac{2\pi z}{l}\right] \tag{4.13}$$

where $\bar{R}$ = average thread radius (m) $= (R_0^2 - \alpha^2/2)^{1/2}$, R_0 = initial thread radius (m), α = distortion amplitude (m), z = coordinate along the thread (m) and l = distortion wavelength (m).

When the distortion wavelength is larger than the circumference of the thread ($l > 2\pi R_0$) the interfacial area decreases as the distortion amplitude increases, i.e. the liquid thread is unstable to distortions. The amplitude α must then grow exponentially with time:

$$\alpha = \alpha_0 e^{qt} \tag{4.14}$$

where α_0 = distortion at $t = 0$ (m), $q = \dfrac{v}{2\eta_2 R_0} \cdot \Omega(l, \lambda)$ = the growth rate (1 s^{-1}),

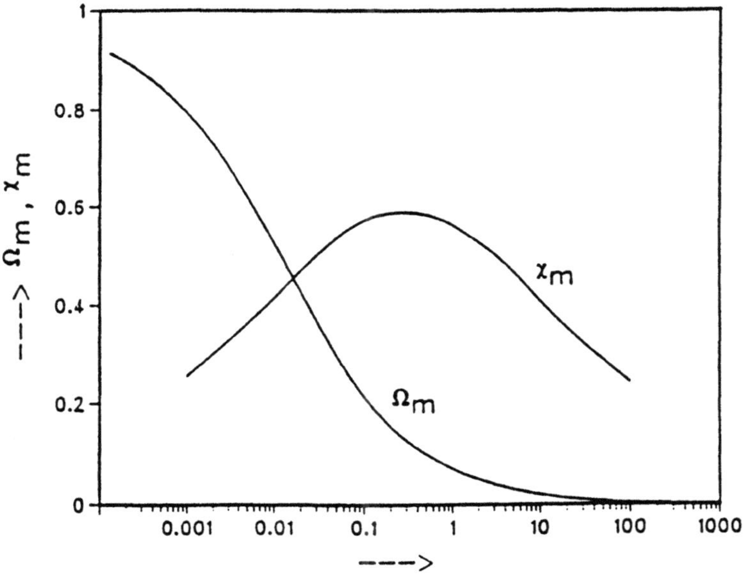

Figure 4.7 Dominant wavenumber X_m and corresponding growth rate Ω_m of interfacial disturbances, as a function of the viscosity ratio λ (with permission from J. Elmendorph [50]).

v = interfacial tension ($N\,m^{-1}$), η_2 = viscosity of the matrix (Pa s), λ = viscosity ratio $= \eta_1/\eta_2$, η_1 = viscosity of the dispersed phase (Pa s) and $\Omega(l, \lambda)$ = a tabulated function [29].

The $\Omega(l, \lambda)$ function shows a maximum for a certain wavelength, l_m. The distortion having this dominant wavelength causes the thread to break up. Figure 4.7 shows the dominant wave number, $X_m = 2\pi R_0/l_m$, and the dominant growth rate, $\Omega(l_m, \lambda)$, as a function of the viscosity ratio λ. Thread break-up is completed when $\alpha = \bar{R}$ and this is the case when $\alpha = 0.81R_0$. From eqn (4.14a) the time for break-up, t_b, can be derived:

$$t_b = \frac{1}{q} \cdot \ln\left[\frac{0.81\,R_0}{\alpha_0}\right] \tag{4.14a}$$

This theoretical value of t_b is only valid when the distortions have sinusoidal and axisymmetric shapes. In practice this is not the case, therefore there is a limited applicability for this formula, namely only when the deformation α is less than, say, $0.6R_0$ or $0.7R_0$. A comprehensive treatment of the rheology of Newtonian colloidal systems can be found in [43].

The description of deformation and break-up given so far is valid for purely viscous Newtonian liquids. Relatively few contributions on drop behaviour in viscoelastic systems have been published [44–49]. Viscoelasticity, and especially a yield stress, may severely slow down or completely prevent the occurrence of thread break-up. Elmendorp [50, 51] showed that in model liquids exhibiting distinct yield stresses no break-up occurs if the yield stress is larger than the pressure difference which is generated in the thread by the different radii of curvature. Verhoogt [52] derived criteria for the break-up of systems with yield stress.

Criteria for drop stability. At infinite dilution in simple shear flow, κ may be expressed as a function of the viscosity ratio, $\kappa_c(\lambda)$. At higher concentration but still discrete drops a dependency of φ is found. In elongational flow, κ (defined as $(\sigma_{xx} \cdot d)/\nu$) is another function of λ ($\kappa_{\text{elong.}} < \kappa_{\text{shear}}$).

Some important consequences of this are:

- Parameters favouring break-up:
  - large droplet size
  - high matrix viscosity
  - small interfacial tension
- In shear flow, droplets with $\lambda > 3$–4 do not break-up (if their viscosity is too high, the drops rotate in the flow).
- Elongational flow (with zero vorticity) is more efficient for droplet break-up (no upper limit for viscosity ratio).
- The domain $0.1 < \lambda < 1$ is most favourable for break-up (shear and elongational flows).
- For low values of λ, droplet elongation at break-up increases strongly.

4.6 The role of coalescence

When the droplet concentration is increased the coalescence must be taken into account. The literature on break-up and coalescence processes in liquid–liquid multiphase systems is extensive. The problem of development of domain size distribution in selected flows has been formulated by Valentas *et al.* [53] in terms of the so-called 'general population balance equation' [54]. Silberberg and Kuhn [55], Tokita [56], Fortelny and Kovár [57], Elmendorph [50] and Elmendorph and Van der Vegt [51] proposed simplified models taking break-up and coalescence into account.

For polymer blends and alloys, solutions of the general population balance equations have not been published. A number of publications deal with the influence of volume fraction of the dispersed phase on polymer blend rheology, drop size and stability, namely, Van Oene [58], Utracki [59], Favis and Chalifoux [60] and Chen *et al.* [61]. The above mentioned topics have been summarised in a number of recent books and reviews [1, 41, 62–66].

A short corollary of the observations is that most theoretical work as well as experimental verification is based on Newtonian drops in Newtonian media. Few contributions deal with viscoelastic drops in viscoelastic or Newtonian media. Exact studies dealing with single drop studies have been performed but only a few involved two body interactions, while multibody interactions and coagulations were treated in a semiempirical approach [43].

Recently Utracki [67] developed an elegant semi-empirical theory which may be used to explain some of the anomalies observed for polymer blends. A brief rendering of Utracki's work follows. Lyngaae-Jørgensen and Utracki [25] reported that for most immiscible polymer blends onset of phase co-continuity occurs near the percolation threshold concentration, $\phi_{cr} = 0.156$. On the other hand, the maximum co-continuity of the phases (or the phase inversion concentration) occurs at a composition in which the viscosity of the system having liquid-1 dispersed in liquid-2 is equal to the viscosity of the system having liquid-2 dispersed in liquid-1. From this postulate came a new relation for the phase inversion concentration, $\phi_{1I} = 1 - \phi_{2I}$:

$$\lambda = [(\phi_m - \phi_{2I})/(\phi_m - \phi_{1I})]^{[\eta]\phi_m} \tag{4.15}$$

where $\lambda = \eta_1/\eta_2$, $[\eta]$, and ϕ_m are, respectively, the viscosity ratio (subscripts 1 and 2 refer to the dispersed and continuous phase, respectively), the intrinsic viscosity of the dispersed phase, and the maximum packing volume fraction. For most polymer alloys and blends $[\eta] \simeq 1.9$ and $\phi_m \simeq 1 - \phi_{cr} \simeq 0.84$. Equation (4.15) was found valid for at least a dozen polymer alloys and blends. In addition, two mechanisms of flow were assumed: (i) the emulsion-like blend behaviour controlled by the relative polymer concentration, ϕ_i/ϕ_{iI} (where ϕ_i is the volume fraction of polymer i in the blend), and (ii) the interlayer slip, the concentration dependence for constant stress viscosity, $\eta = \eta(\sigma)$, was expressed as:

$$\log \eta = \log \eta_L + \Delta \log \eta^E \tag{4.16}$$

where

$$\log \eta_L = -\log[1 + \beta(\phi_1 \phi_2)^{1/2}] - \log(\phi_1/\eta_1 + \phi_2/\eta_2)$$

and

$$\Delta \log \eta^E = \eta_{max}\{1 - [(\phi_1 - \phi_{1I}^2/(\phi_1 \phi_{2I}^2 + \phi_1 \phi_{1I}^2)]\}$$

where $\beta = \beta(\sigma_{12})$, the interlayer slip factor, and $\Delta \log \eta^E$ is an excess term derived from the concept of the emulsion-like behaviour of polymer blends.

The expression given by eqn (4.16) can be used to explain many of the observed viscosity–volume fraction relationships observed for polymer blends. Maxima in such relations are due to the second term on the right side of eqn (4.16). Minima may be explained by interfacial slippage and expressed through the term $\log \eta_L$.

4.7 Non-steady-state flow (transition to steady-state)

4.7.1 The role of capillarity number and stability time

As documented above the deformational behaviour of a single drop may be described by two dimensionless groups: the capillarity number which for simple shear flow is defined by $\kappa \equiv \sigma_{12}\, d/\nu$; and the viscosity ratio ($\lambda \equiv \eta_1/\eta_2$), where η_1 indicates the viscosity of the dispersed phase, η_2 is the viscosity of the matrix phase, and σ_{12}, d and ν are shear stress, droplet diameter and interfacial tension coefficient. In simple shear flow a necessary condition for break-up is $\lambda \leqslant 4$.

If the capillarity number exceeds its critical value for drop break-up, $\kappa_{cr} \simeq 1$, the viscous shear stress becomes larger than the interfacial stress, the drop is extended and finally breaks up into smaller droplets. If $\kappa < \kappa_{cr}$ the interfacial stress can equilibrate the shear stress and the drop will only deform into a stable prolate ellipsoid.

A subdivision of the dispersion processes may be based on the value of the capillarity number:

(i) Distributive flow processes when $\kappa \gg \kappa_{cr}$ (large dispersed domains): drops extend affinely with the matrix into long fibrillae, but do not develop capillarity waves leading to break-up, since the interfacial stress is overruled by the shear stress. The fibrillae break into drops only when the flow is stopped.

(ii) Dispersive flow processes when $\kappa \geqslant \kappa_{cr}$ (locally small radii of curvature): v/d now competes with σ_{12} and causes disturbances at the interface to grow, leading to break-up during the flow into smaller droplets and thus a finer dispersion.

Thus, for shear flow of dilute Newtonian emulsions:

κ-value	*Drop deformation*
$\kappa < 0.1$	no deformation
$0.1 < \kappa < 1$	prolate ellipsoids at equilibrium
$1 \leqslant \kappa \leqslant 4$	break into two principal plus satellite drops
$4 < \kappa$	affine deformation into long fibre

Therefore, for $\kappa \gg \kappa_{cr}$ a drop is unstable because the shear forces are much larger than the interfacial stress forces trying to keep it together. Consequently a drop deforms to very elongated forms and eventually breaks. Large drops (millimetre sized) will deform affinely with the matrix. For a spherical drop of diameter a deformed to an ellipsoid with length L and width B for total shear $\gamma \geqslant 5$ (and $\kappa \gg \kappa_{cr}$) [65]:

$$L/a \simeq \gamma$$

$$B/a \simeq \gamma^{-1/2}$$

$$D_c \cong \frac{\gamma^{3/2} - 1}{\gamma^{3/2} + 1}$$

for simple shear flow.

In elongational flow, the length of an affinely deforming drop increases exponentially:

$$L/a = e^{\varepsilon}$$

$$B/a = e^{-\varepsilon/2}$$

$$D = \frac{e^{3\varepsilon/2} - 1}{e^{3\varepsilon/2} + 1}$$

Only the total strain is important: $\varepsilon = \dot{\varepsilon}t$, the product of the elongation rate $\dot{\varepsilon}$ and time. In elongational flow L/a and D increase much faster than in simple shear.

An elongating drop remains oriented in the direction of extensional deformation during simple elongation. This contrasts with the drop behaviour in simple shear flow where the drop rotates.

A dimensionless break-up time can be defined as:

$$t_b^* = t_b \frac{\nu}{\eta_2 R}$$

Grace [42] evaluated t_b^* as a function of λ for viscoelastic drops in viscoelastic matrices. Results are given in Figure 4.8. If the residence time in a given process is less than the break-up time for elongated species, break-up will not take place.

4.7.2 *Structure development during simple shear flow of polymer blends*

This section deals with dispersive flow processes at $\kappa \geqslant 1$. Let us consider a single drop in a continuum where $\eta_1 < 4\eta_2$. For $\kappa \geqslant \kappa_{cr}$ it will break when the flow time exceeds the break-up time t_b. For $\kappa < \kappa_{cr}$ the drop will be stable or, stated differently, the break-up rate, R_b, will be zero. Thus

$$R_b = R_b(\kappa, t/t_b) \qquad \text{for } \kappa \geqslant \kappa_{cr}$$
$$R_b = 0 \qquad \text{for } \kappa < \kappa_{cr}$$

However, at finite volume fraction φ of the minor phase the coalescence occurs at a rate $R_c = R_c(\varphi, \kappa, \lambda)$ at all κ values; in steady-state $R_b = R_c$ for all volume fractions. This means that *the average steady-state drop size* $\bar{\mathrm{d}}$ *found in a blend*

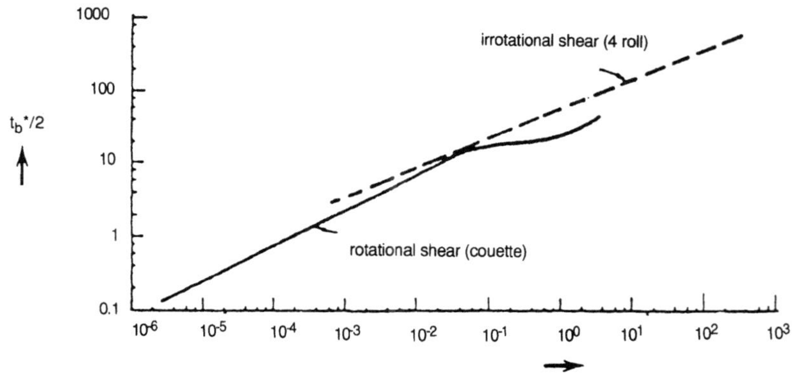

Figure 4.8 Dimensionless time for drop break up at κ_{cr} as function of the viscosity ratio (Grace, [42]) (reprinted with permission of Gordon and Breach Publishers Inc.).

at finite volume fraction φ *is different from the maximum drop size* d *predicted for* $\kappa = \kappa_{cr}$ *for a single drop*. Only for $R_c^{\text{steady-state}} \to 0$ do these two diameters coincide: $\bar{d} \to d$.

The formation of fibre-like drops in simple shear flow has recently been investigated with a model system consisting of polystyrene and polymethyl methacrylate [68, 69]. A model has been developed based on the following ideas.

When discrete domains of a well-dispersed minor phase of an immiscible blend are deformed in a simple shear flow, elongated ellipsoidal domains may form. Their disintegration and coalescence determine the morphology. In a system with capillarity number, $\kappa \geqslant 1.0$, and viscosity ratio, $\lambda \ll 1$, at finite concentration of the minor phase, viscoelastic liquid drops in a viscoelastic liquid matrix tend to form long ellipsoidal, nearly cylindrical, domains. These elongated domains in turn disintegrate into smaller drops. In the proposed description of drop elongation and break-up it is assumed that the total domain size distribution can be divided into small spherical and large elongated drops. Simulation of the number average aspect ratio, volume fraction of large domains, and semi-axis length of the ellipsoids are accomplished by the simultaneous solution of three rate equations. These are the rate of change of the number of large and small drops, and the rate of change of the number average small semiaxis. The model fits the experimental aspect ratio versus time dependence for polystyrene–polymethylmethacrylate blends. The steady-state data show a maximum in the number average aspect ratio near the transition from Newtonian to pseudoplastic flow. The difference in normal stress between the matrix and the dispersed phase seems to constitute the principal deforming force in the non-Newtonian range or, stated differently, a necessary condition for formation of fibre-like structures in the non-Newtonian range is that the first normal stress difference of the matrix phase is larger than that of the minor phase at the same shear stress.

4.8 Complex flow: processing of polymer blends

In this section the processing of thermoplastic materials is considered. The processing of polymer blends in, for example, extruders and injection moulding equipment may be divided into zones such as feeding, solid transport and melting (zone 1), melt mixing (zone 2), flow of polymer melts (zone 3) and solidification (zone 4). Each zone represents characteristic problems from the point of view of morphology development [70, 71].

The development of microstructure starts when the individual blend components are brought into physical contact and ends when the structure is frozen-in during the shaping operations.

At the beginning of the processing operation the scaling in the mixture may be given by the domain size distribution of the minor component, which is typically several millimetres in magnitude. During melting the scale of segregation may easily be reduced by orders of magnitude down to, say, 50 μm. Processing through zones 2 and 3 reduces the domain size to a final dispersed state with average size typically below 1 μm.

A total simulation of the behaviour of polymer blends in all four zones is necessary for detailed production of immiscible polymer blends. Extensive research is aimed at such simulation but this will probably not be achieved in the near future.

At present similitude studies and dimensional analyses seem to give as good (or as poor) guidance as more rigid methods. The case of stable drops at steady state and infinite dilution in simple shear flow is an example of the usefulness and simplicity of dimensional analysis. The difficult step in dimensional analysis is the first one: establishment of a relation between a variable and the necessary independent variables. A literature study gives (of course) that the shear stress may be expressed as

$$\sigma = F(d, \eta_1, \eta_2, v)$$

A dimensional analysis (see e.g. [90]) gives immediately that two independent dimensionless groups are sufficient: η_1/η_2 and $\sigma \cdot d/v$.

4.8.1 Flow through a contraction: fibre formation

Practical extrusion processes often involve geometrically complex dies. Such dies are usually tapered, or streamlined, to achieve maximum output rate under conditions of laminar flow. These converging flows may be analysed in terms of their extensional and simple shear components to calculate the relationships between volume flow rate, pressure drop, and post-extrusion swelling. The analysis can also be extended to cover the free convergence as fluid flows from a reservoir into a die [72–74].

In most cases the extensional flow component dominates the flow in the entrance region to a capillary die. A number of studies of flow in capillary rheometers may be treated in this connection (Figure 4.9).

Elongational flow may lead to fibre formation [64]. According to Elmendorf and Van der Vegt [51], break-up of droplets in extensional flow is not to be expected. The continuous elongation of the droplet does not allow the instability causing break-up to grow to such an extent that actual burst into two halves can be completed. As time-independent extensional flow fields are not likely to be encountered in actual polymer blending devices, the break-up of droplets due to extensional flow fields can only occur by capillary break-up of highly elongated droplets after conical or wedge shaped contractions. Thus flow in capillary rheometry should be divided into flow in the inlet zone and flow in the capillary. As shown on the principal sketch on Figure 4.9, which was proposed by Vinogradov *et al.* [75], a well-dispersed blend will be deformed to a fibrillar structure in the extensional flow dominated inlet zone

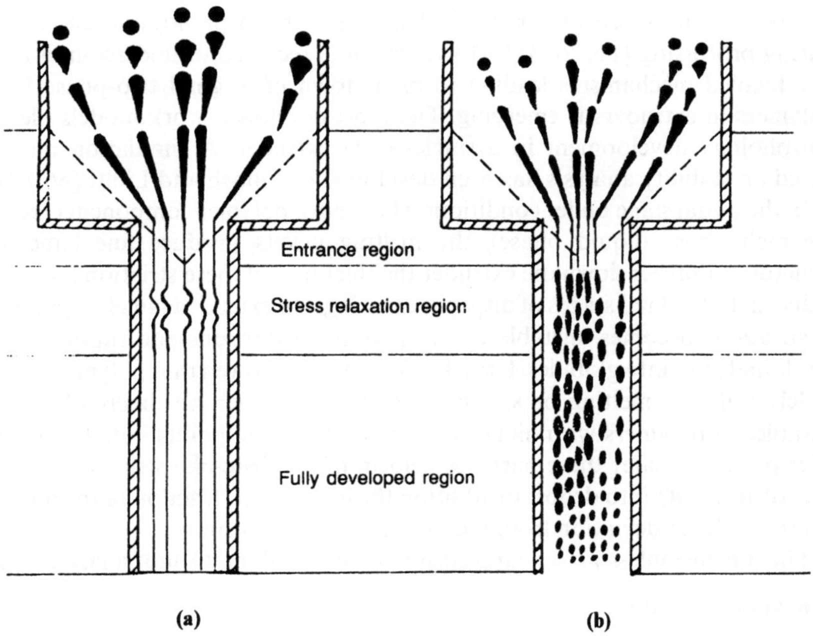

Figure 4.9 Empirical model representation of molten two-phase dispersed polymer blend flowing through a capillary.

and may break up in the capillary if the residence time in the capillary is longer than the break-up time for the fibre-like domains.

As an example, the residence time in a circular capillary with length L and radius R may be estimated as

$$t_{\text{res.}} = \frac{L\pi R^2}{Q}$$

where Q is the volumetric flow rate and compared with the empirical break-up time read from Figure 8.4. Well-dispersed polymer blends are considered in this chapter, that is $d/R \ll 1$ where d is the diameter of the drops before entering the inlet zone to the capillary and R is the capillary radius.

The lesson to be learnt from this case is that if long fibres are required, the capillary part of the die should be avoided and the extrudate should be drawn directly from the inlet zone. Furthermore a small viscosity ratio $\lambda = \eta_1/\eta_2$ should give longer fibre-like domains.

4.8.2 Breakdown of pellets in extruders and mixing equipment

The major morphological changes which takes place during processing from pellet size to submicron size often occur during the initial softening stage. A number of papers have been published dealing with morphology development during processing [65, 76–85]. However, quite recently an understanding of the detailed mechanisms leading to breakdown of original two-phase feed polymers in extruders is emerging. The most developed work models blend morphology development in a single-screw extruder. A simulation model based on stability analysis has been developed by Ghosh and Lindt [86, 87].

In the initial stage under conditions where both polymer components are in the melt phase (liquid phase), the melting pellets produce fine lamellar structures. Further down the extruder the thickness of these striation layers is reduced. In the final stages of dispersive mixing, the extended lamellar pattern of striations becomes unstable due to periodic distortions and rupture into fibrils and later into globules. Utracki and Shi [88] showed that polymers with widely different melting or softening temperatures behave differently. The granules of polymers with higher melting or softening points initially act as filler particles. Later these particles deform into a fibre-like structure which may disintegrate into drops. In addition the low volume fraction of the minor phase results in deviation from the multilayer mechanism.

The dimensionless groups appearing in the Ghosh and Lindt analysis are:

- viscosity ratio λ
- capillarity number $\kappa \equiv \dfrac{\sigma \cdot h}{v}$
- density ratio
- thickness ratio h/H
- Reynolds number $\dfrac{\rho v h}{\eta_2}$
- dimensionless time

where H is the total thickness of molten layer between solid surface and solid polymer bed and h is an average thickness of the multilayer structure. Some general guidelines are:

- In the early melting stages a stable array of striations is formed if both polymers are liquid and not at low volume fraction.
- Once below a critical characteristic thickness, break-up of the layers begins in a preferential direction determined by the relative rates of the disturbances.

Sundararaj *et al.* [89] studied morphology development in the initial softening stage in a co-rotating twin-screw extruder and compared the results with initial morphology in a batch mixer. Sheet break-up studies under quiescent conditions were also performed. The authors conclude that formation of thin layers did occur both in twin-screw extruders and in the batch mixer. The layer-forming mechanism causes the phase size to decrease tremendously during the melting and softening stage. Most of the significant morphology development occurs immediately after the first point of melting in the extruder and within the first minute in the batch mixer. The break-up mechanism proposed involves hole formation in the layers as a result of impurities in the blend. The holes expand due to interfacial tension forces until coalescence into thin ligaments which successively break up into small drops.

4.8.3 Guidelines for blends morphology after processing

Utracki and Shi [66] give the following advice: 'It is evident that the literature information on the microrheology of liquid/liquid systems can be of little direct use for predicting the development of polymer blend morphology during compounding in twin-screw extruders. However, the following general principles can serve as guides:

- Drops with $\lambda \geqslant 3.5$ cannot be dispersed in shear but can be so in extension.
- The larger the interfacial tension coefficient, the less the droplet will deform.
- t_b^* and κ_{cr} are two important parameters describing the break-up process; $\kappa > \kappa_{cr}$ is the principal requirement for drop break-up but the necessary time, $t \geqslant t_b$, must be provided to complete the process.
- Effects of coalescence must be considered, even for relatively low concentration of the dispersed phase $\phi_d \geqslant 0.005$.'

It seems clear that a successful simulation of, say, mixing extruders demands valid modelling of the (four) extruder zones mentioned in section 4.8. The present knowledge is insufficient but the following simple approach may be applied as a first approximation. The procedure is applied to a given processing unit.

1. Evaluate the average residence time, t_r, in the unit.
2. Compare the residence time with the break-up time, t_b, estimated as described in section 4.7.1 (Figure 4.8).
3. If $t_r \gg t_b$ and $\lambda < 3.5$ it can be assumed that dispersive flow can be obtained.

4a. If nearly spherical discrete drops are obtained after processing for $\varphi \lesssim \varphi_c$ a plot of capillarity number against viscosity ratio valid for a given processing unit can be constructed. Wu [79] did successfully apply such an approach with a co-rotating twin-screw extruder as shown in Figures 4.10 and 4.11.

4b. If discrete cylinder-like ellipsoids are obtained after processing the die section, including the inlet, can be analysed separately from the rest of the processing unit.

Assuming that nearly spherical domains are obtained before the die section, the procedure given in 4a can be used on the blend state before the die section. The die section may then be studied (and analysed) separately as described in section 4.8.1.

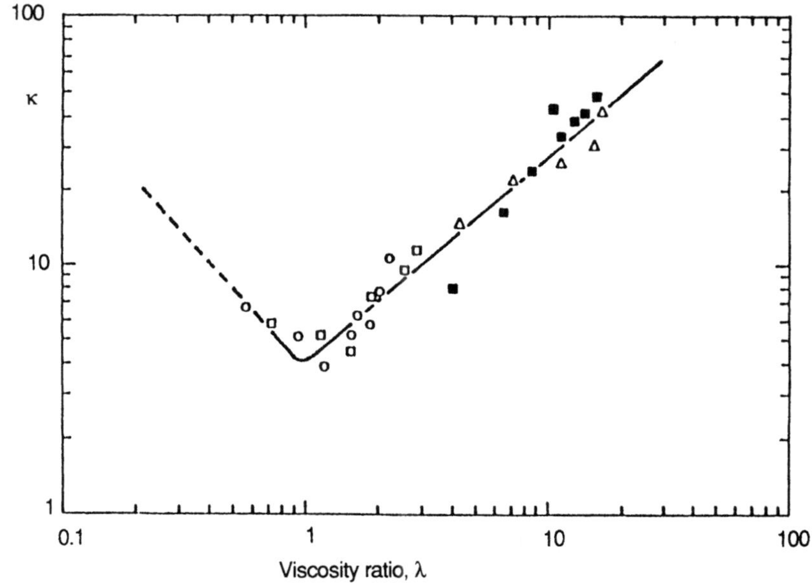

Figure 4.10 Dimensionless master curve of ($\dot{\gamma}\eta_2 a_n/v$) versus η_1/η_2 for the melt extrusion using a co-rotating twin-screw extruder [79]. ○, PET/EP rubbers; ■, nylon Z-1/EP rubbers; □, nylon Z-2/EP rubbers; △, nylon Z-1/EPX rubbers (reprinted with permission of Society of Plastics Engineers).

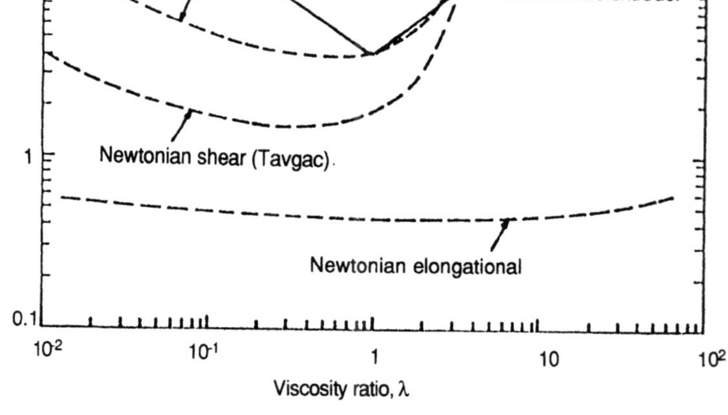

Figure 4.11 Comparison of master curves for viscoelastic blends in the co-rotating twin-screw extruder against the Newtonian systems in steady and uniform shear field and in steady and uniform elongational field [79] (reprinted with permission of Society of Plastics Engineers).

Nomenclature

B	minor axis of ellipsoid
d	droplet diameter
D	drop deformation, $(L-B)/(L+B)$
$f(x)$	a function of x
ΔG_m	changes in Gibb's free energy on mixing
G^*, G', G''	complex modulus, storage modulus and loss modulus
$H = M_w/M_n$	polydispersity factor
ΔH_m	change in enthalpy on mixing
L	major axis of ellipsoid
L	capillary length
l	distortion wavelength
M_n, M_w	number and weight average molecular weight
$N_{1,i}$	first normal stress difference for the dispersed ($i = 1$) and the matrix ($i = 2$) phase
PMMA	polymethylmethacrylate
PS	polystyrene
q	growth rate of distortion
δ	unit tensor
SAN	copolymer of acrylonitrile and styrene
S_{xx}	tensile stress
ΔS_m	entropy change by mixing
SEC	size exclusion chromatography
t	time
t_b	time necessary for break-up of droplet
T	temperature
TEM	transmission electron microscopy
X_m	dominant wave number
$\bar{X}_n$	average number of repeat units per polymer chain
v, v_x, v_y, v_z	velocity, velocity in the flow, the gradient and the neutral direction, respectively
V_x	velocity of the upper plane of two parallel planes
δ	phase angle
$\gamma, \gamma^*, \gamma', \gamma''$	shear strain (dx/dy), complex shear strain, real part of shear strain, imaginary part of shear strain
$\dot{\gamma}$	shear rate
$\dot{\gamma}_E$	elongational rate
η, η_0	viscosity, zero-shear viscosity
η^*, η', η''	complex viscosity, in phase dynamic viscosity and out of phase dynamic viscosity
η_1, η_2	viscosity of dispersed and matrix phase, respectively
η_E	elongational viscosity
λ	viscosity ratio

κ	capillarity number
κ_{cr}	critical capillarity number
$\sigma, \sigma_{x,y}$	shear stress
σ_c	critical shear stress
$\sigma_{xx} - \sigma_{yy} = N_1$	the first normal stress difference
σ	stress tensor
σ_{ij}	stress tensor components
ν	interfacial tension
φ	volume fraction of dispersed phase
φ_2	volume fraction of phase 2
$\psi_1(\dot{\gamma})$	first normal stress coefficient
$\psi_2(\dot{\gamma})$	second normal stress coefficient
ρ, ρ_d	density, droplet density
θ	scattering angle
$\Omega(l, \lambda)$	a tabulated function
ω	circular frequency

Subscripts

1, 2	dispersed phase, matrix
0	equilibrium value

References

1. Utracki, L.A. (1990) *Polymer Alloys and Blends, Thermodynamics and Rheology*, Hanser Publications.
2. Lyngaae-Jørgensen, J. (1985) Melt flow properties of block copolymers, *Processing, Structure and Properties of Block Copolymers*, Folkes, M.J., ed., Elsevier Applied Science Publishers, London, Chapter 3, pp. 75–123.
3. Bird, R.B., Armstrong, R.C. and Hassager, O. (1987) *Dynamics of Polymeric Liquids*, Wiley.
4. Tanner, R.I. (1985) *Engineering Rheology*, Oxford Science Publishers.
5. Vinogradov, G.V. and Ya Malkin, A. (1980) *Rheology of Polymers*, Springer-Verlag, Berlin.
6. Walters, K. (ed.) (1982) *Rheometry: Industrial Applications*, Research Studies Press, Chichester.
7. Dealey, J.M. (1982) *Rheometers for Molten Plastics*, Van Nostrand Reinhold, New York.
8. Denson, C.D. (1973) *Polym. Eng. Sci.*, **13**, 125.
9. Lyngaae-Jørgensen, J. (1989) Phase transitions in simple flow fields, *ACS Symposium Series No. 395*, ACS Washington DC, Chapter 6.
10. Lyngaae-Jørgensen, J. and Søndergaard, K. (1987) *Polym. Eng. Sci.* **27**, 351.
11. Lyngaae-Jørgensen, J. (1988) *Proceedings ACS, DIV PMSE* **58**, 702.
12. Lyngaae-Jørgensen, J. and Søndergaard, K. (1987) *Polym. Eng. Sci.* **27**, 344.
13. Silberberg, A. and Kuhn, W. (1952) *Nature* **170**, 450.
14. Cheikh Larbi, F.B., Malone, M.F., Winter, H.H., Halary, J.L., Leviet, M.H. and Monnerie, L. (1988) *Macromolecules* **21** 3532.
15. Kammer, H.W., Kummerloewe, C., Kressler, J. and Melior, J.P. (1991) *Polymer*, **32**(8), 1488.
16. Hashimoto, T., Takebe, T. and Suehiro, S. (1988) *J. Chem. Phys.* **88**(9), 5874.
17. Katsaros, J.D., Malone, M.F. and Winter, H.H. (1989) *Polym. Eng. Sci.* **29**(20), 1434.
18. Hindawi, I.A., Higgins, J.S. and Fernandex, M.L. (1991) *PPS Meeting, Palermo*, September 15-18, Abstracts pp. 169-170.

19. Mazich, K.A. and Carr, S.H. (1983) *J. Appl. Phys.* **54**(10), 5511.
20. Hindawi, I., Higgins, J.S., Galambos, A.F. and Weiss, R.A. (1990) *Macromolecules* **23**(2), 670.
21. Rector, L.P., Mazich, K.A. and Carr, S.H. (1988) *J. Macromol. Sci. Phys.* **327**(4), 421.
22. Katsaros, J.D., Malone, M.F. and Winter, H.H. (1986) *Polymer Bulletin* **16**, 83.
23. Nakatani, A.I., Kim, H. and Han, C.C. (1990) *Mater. Res. Soc. Symp. Proc.* **166**, 479.
24. Mani, S., Malone, M.F., Winter, H.H., Halary, J.L. and Monnerie, L. (1991) *Macromolecules*, **24**, 5451.
25. Lyngaae-Jørgensen, J. and Utracki, L.A. (1991) *Makromol. Chem., Macromol. Symp.* **48/49**, 189.
26. Lyngaae-Jørgensen, J., Kuta, A., Søndergaard, K. and Venø Poulsen, K. Structure and properties of polymer blends with dual continuity. *Polym. Networks and Blends* (in prep.).
27. Taylor, G.I. (1934) *Proc. Roy. Soc. London* **A146**, 501.
28. Taylor, G.I. (1932) *Proc. Roy. Soc. London* **A138**, 41.
29. Tomotika, S. (1935) *Proc. Roy. Soc. London* **A150**, 322.
30. Tomotika, S. (1936) *Proc. Roy. Soc. London* **A153**, 302.
31. Acrivos, A. and Lo, T.S. (1978) *J. Fluid. Mech.* **86**, 641.
32. Rallison, J.M. and Acrivos, A. (1978) *J. Fluid Mech.* **89**, 91.
33. Karam, H. and Bellinger, J.C. (1968) *Ind. Eng. Chem. Fundam.* **7**, 576.
34. Gordon, M., Yerushalmi, J. and Shinner, R. (1973) *Trans. Soc. Rheol.* **17**, 303.
35. Goldsmith, H.L. and Mason, S.G. (1967) in *Rheology: Theory and Applications*, Vol. 4, Eirich, F.R., ed., Academic Press, New York.
36. Mikami, T., Cos, R.G. and Mason, S.G. (1975) *Int. J. Multiphase Flow* **2**, 113.
37. Rumscheidt, F.D. and Mason, S.G. (1961) *J. Colloid Sci.* **16**, 210.
38. Rumscheidt, F.D. and Mason, S.G. (1962) *J. Colloid Sci.* **17**, 260.
39. Torza, S., Cox, R.G. and Mason, S.G. (1992) *J. Colloid Interface Sci.* **38**, 395.
40. Tjahjadi, M. and Ottino, J.M. (1991) *J. Fluid. Mech.* **232**, 191.
41. Han, C.D. (1981) *Multiphase Flow in Polymer Processing*, Academic Press, New York.
42. Grace, H.P. (1981) *Eng. Found. Res. Conf. Mixing*, Academic Press, New York.
43. Van De Ven, Theo. G.M. (1989) *Colloidal Hydrodynamics*, Academic Press.
44. Choi, S.J. and Schowalter, W.R. (1975) *Phys. Fluids* **18**, 420.
45. Flumerfelt, R.W. (1972) *Ind. Eng. Chem. Fundam.* **11**, 312.
46. Flumerfelt, R.W. (1980) *J. Colloid Interface Sci.* **76**, 330.
47. Chin, H.B. and Han, C.D. (1980) *J. Rheol.* **24**, 1.
48. Lee, W.-K., Yu, K.-L. and Flumerfelt, R.W. (1981) *Int. J. Multiphase Flow* **7**, 385.
49. Lee, W.-K. and Flumerfelt, R.W. (1981) *Int. J. Multiphase Flow*, **7**, 363.
50. Elmendorph, J.J. (1986) *A Study on Polymer Blending Microrheology*, Ph.D. Thesis, Delft.
51. Elmendorph, J.J. and van der Vegt, A.K. (1991) Two-phase polymer systems, in *Progress in Polymer Processing*, Vol. 2, Utracki, L.A., ed.
52. Verhoogt, H. (1992) *Morphology, Properties and Stability of Thermoplastic Polymer Blends*, Ph.D. Thesis, Delft.
53. Valentas, K.J., Bilous, Q. and Amundson, N.R. (1966) *Ind. Eng. Chem. Fundam.* **5**, 271.
54. Coulaloglou, C.A. and Taularides, L.L. (1977) *Chem. Eng. Sci.* **32**, 1289.
55. Silberberg, A. and Kuhn, W. (1954) *J. Polym. Sci.* **13**, 21.
56. Tokita, N. (1979) *Chem. Technol.*, **50**, 292.
57. Fortelny, I. and Kovár, J. (1989) *Eur. Polym. J.*
58. van Oene, H. (1978) in *Polymer Blends*, Vol. 1, Paul, D.R. and Newman, S., eds., Academic Press, New York, Chapter 7.
59. Utracki, L.A. (1983) *Polym. Eng. Sci.* **23**, 602.
60. Favis, B.D. and Chalifoux, J.P (1988) *Polym. Eng. Sci.* **28**, 69.
61. Chen, C.C., Fontan, E., Min, K. and White, J.L. (1988) *Polym. Eng. Sci.* **28**, 69.
62. Utracki, L.A. (1987) in *Current Topics in Polymer Science*, Vol. II. Ottenbrite, R.M., Utracki, L.A. and Unie, S., eds., Hanser Publications, New York.
63. Wu, S. (1987) *Polym. Eng. Sci.* **27**, 335.
64. Tsebrenko, M.V., Danilova, G.P. and Malkin, A. Ya. (1989) *J. Non-Newtonian Fluid Mech.* **31**, 1.
65. Meijer, H.E.H. and Janssen, J.M.H. (1991) Mixing of immiscible liquids, WFW report. Eindhoven University of Technology (to be published in *Mixing and Compounding—Theory and Practical Progress*, Vol. 4, in Progress in Polymer Processing Series).

66. Utracki, L.A. and Shi, Z.H. (1992) *Polym. Eng. Sci.* **32**(24), 1824.
67. Utracki, L.A. (1991) *J. Rheol.* **35**, 1615.
68. Lyngaae-Jørgensen, J. and Valenza, A. (1990) *Makromol. Chem., Macromol. Sym.* **43**.
69. Lyngaae-Jørgensen, J., Søndergaard, K., Utracki, L.A. and Valenza, A. (1993) "Formation of Ellipsoidal Drops in Simple Shear Flow, Transient Behaviour". *Polymer Networks and Blends* (in prep.).
70. Utracki, L.A. and Lindt, T. to be published.
71. White, J.L. and Min, K. (1988) *Makromol. Chem., Macromol. Symp.* **16**, 19.
72. Cogswell, F.N. (1972) *Polym. Eng. Sci.* **12**, 64.
73. Tremblay, B. (1989) *J. Non-Newtonian Fluid Mech.* **137**.
74. Han, C.D. and Funatsu, K. (1978) *J. Rheol.* **22**, 113.
75. Vinogradov, G.V., Yokob, G.V.M., Tsebrenko, M.V. and Yudin, A.V. (1974) *Int. J. Polym. Metals* **3**, 99.
76. Schreiber, H.P. and Olguin, A. (1983) *Polym. Eng. Sci.* **23**, 129.
77. Karger-Kocsis, J., Kalló, A. and Kuleznev, V.N. (1978) *Polymer* **19**, 448.
78. White, J.L. and Min, K. (1985) *Adv. Polym. Technol.* **5**, 225.
79. Wu, S. (1987) *Polym. Eng. Sci.* **27**, 335.
80. Heikens, D. and Bartensen, W. (1977) *Polymer* **18**, 69.
81. Heikens, D., Hoen, N., Bartensen, W., Piet, P. and Landan, H.J. (1978) *Polym. Sci., Polym. Symp.* **62**, 309.
82. Elemans, P.H.M. (1989) Ph.D. Thesis, Eindhoven University of Technology, Netherlands.
83. Plochocki, A.P., Dagli, S.S. and Andrews, R.D. (1990) *Polym. Eng. Sci.* **30**, 741.
84. Favis, B.D. (1990) *J. Appl. Polym. Sci.* **39**, 285.
85. Scott, C.E. and Macosko, C.W. (1991) *Polym. Bull.* **26**, 341.
86. Lindt, J.T. and Ghosh, A.K. (1992) *Polym. Eng. Sci.* **32**, 1802.
87. Ghosh, A.K., Ranganathan, S., Lorek, S. and Lindt, J.T. (1991) *ANTEC*, Montreal, Canada, pp. 232–236.
88. Utracki, L.A. and Shi, Z.-H. (1992) *Polymer Processing Society*, PPS-8, New Delphi, 24–27 March, Abstracts pp. 232–234.
89. Sundararaj, U., Macosko, C.W., Rolando, R.J. and Chan, H.T. (1992) *Polym. Eng. Sci.* **32**, 1814.
90. Fox, R.W. and McDonald, A.T. (1973) *Introduction to Fluid Mechanics*, J. Wiley & Sons, New York.

5 Practical techniques for studying blend microstructure

D. VESELY

This chapter describes some methods which are used to characterise the microstructure of polymer blends. The main interests of those who are developing new polymer blends is to investigate and measure the degree of miscibility and the interaction of phases, as they can significantly influence the properties and behaviour of the compound. Phase size and size distribution can be revealed directly by microscopy but to obtain information on partial miscibility and interfacial interaction is more difficult and microscopy has to be supported by other techniques for the full characterisation of a polymer blend. All the techniques can provide some useful information in capable hands, but also frustration and wrong results for the less skilled. The preference for a particular technique is dependent on the polymer system, availability of instruments, experience and personal choice. For this reason this account is naturally biased and represents only one possible approach. There is an extensive literature on characterisation of polymer blends, indicating the importance of the subject (for review see e.g. [1–2]).

5.1 Light microscopy

Light microscopy is one of the most important techniques in the study of polymer blends, as it is relatively simple and cheap and the specimen preparation is not too complicated. However, the resolution of a light microscope is limited by the wavelength of visible light and by the quality of the objective lens. For the best lenses the resolution can be as low as 0.2 μm, which is sufficient to study phases in most immiscible blends but is not sufficient for small phases in toughened, filled or partially miscible polymer blends. The visibility of phases can be enhanced by contrast techniques and the phases can be identified by staining or fluorescence. The main problem with light microscopy, as with other microscopical techniques, is the specimen preparation, as most techniques affect the microstructure. A good review of polymer light microscopy is in [3–4].

5.1.1 The microscope

The principle of a light microscope is shown diagrammatically in Figure 5.1. The light source must be sufficiently powerful to provide good illumination at high magnifications and to reduce the exposure time to less than one minute. The light is condensed by a condenser onto the specimen. The cone of light at this point must match the cone of light entering the objective lens. This is expressed by a numerical aperture and is written on the barrel of the objective lens body. For even illumination, Koehler illumination must be obtained. This requires focusing the condenser in such a way that the image of the filament coincides with the back focal plane of the objective, or the field aperture is

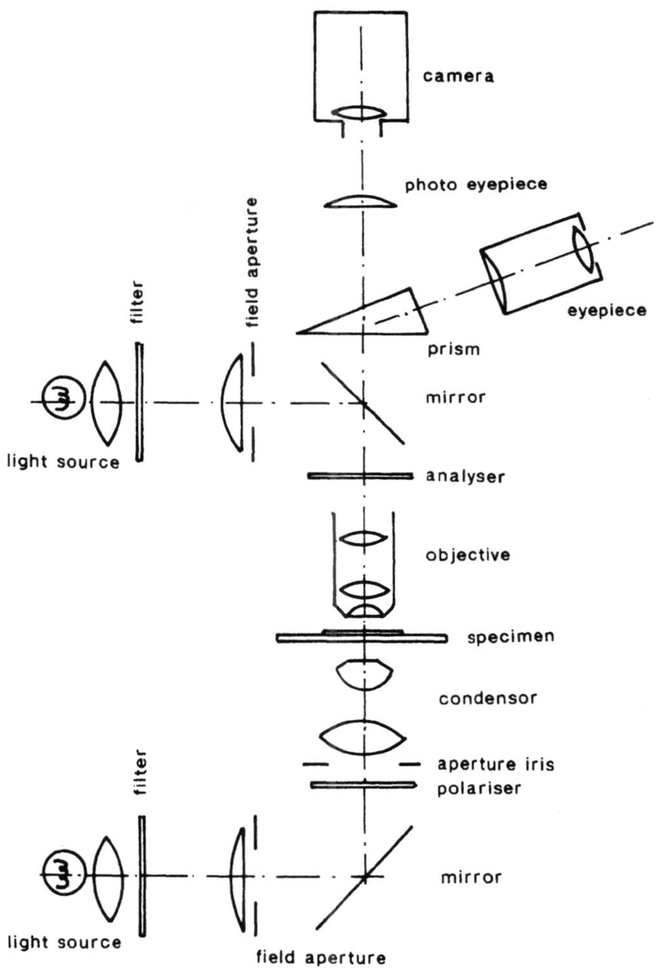

Figure 5.1 Schematic diagram of a light microscope.

focused on the specimen image. For incident light illumination, the light source is placed above the objective and light is reflected through a semitransparent (and often annular) mirror into the objective lens which assumes the dual role of the condenser and the objective.

Aberrations are the major problem of light microscope lenses and their quality (and therefore their cost) is dependent mainly on the degree of correction. Chromatic aberration, planarity of the field and internal stress fields are the major imperfections. When these are eliminated, spherical aberration and astigmatism are usually also well corrected. For better resolution (and thus higher magnification), better quality lenses are required.

The eyepiece must be focused on the latent image produced by the objective lens at the primary image plane. The eyepiece often has a graticule or a cross-wire at this plane to assist in focusing. Because some optical corrections are completed in the eyepiece, it is essential to use the correct combination of lenses specified by the manufacturer. More expensive eyepieces have a wider field of view and a focusing point well above the final lens for use with spectacles. Magnification of the eyepieces is 5 ×, 10 ×, or 12 ×. Higher magnifications are not used, as the light intensity is diminished.

For low magnification (up to 50 ×) a stereo microscope can be used. In principle this consists of two microscopes, one for each eye, so that stereo vision is obtained. It can be used in both transmission and reflection and with polarised light. This microscope is an essential extension of the naked eye vision and is used for the first assessment of features to be studied, for evaluation of the sample quality and imperfections, and for micromanipulation.

5.1.2 Contrast enhancement techniques

1. *Polarised light.* Polarised light can significantly enhance contrast in anisotropic (crystalline or oriented) transparent materials and is often used in conjunction with other types of contrast. It can also be used for the identification of compounds. Polarised light passing through an anisotropic object is rotated due to the different velocities of the ordinary and the extraordinary beam. The amount of rotation is dependent on the orientation of the principle axis and on the specimen thickness. This rotation of polarised light can be utilised for contrast or colour differentiation between different regions or phases in a polymer blend.

Light is polarised by a polarising filter, placed below the condenser and analysed by an analyser, which is placed above the objective lens. At least one of these filters must be rotatable to 90° and for serious work a rotatable specimen stage is also required. The strong contrast from stressed regions of polymer (e.g. produced by cutting) makes it difficult to identify fine phase dispersion and therefore this contrast technique is not ideal for the study of blends containing crystalline polymers.

2. *Phase contrast.* Phase contrast is the most valuable contrast enhancement for amorphous materials as it is sensitive to small changes in the refractive indices. It will reduce the resolution substantially on thicker or high contrast samples and is thus most suitable for low contrast objects in very thin samples. It cannot be used reliably for identification of compounds.

The technique requires an annular aperture in the condenser (cone illumination) and a phase plate of equivalent diameter and shape in the back focal plane of the objective lens. This means that a special objective lens (PHACO) is needed. Very high contrast is obtained (Figure 5.2a) and this technique is therefore the most important technique for the study of polymer blends.

3. *Differential interference contrast.* This technique was originally designed for incident light and for the study of lightly etched surfaces or surface oxides. It is very sensitive to surface steps (steps 5 nm high can be resolved). It can also be used for transmission microscopy, as it is very sensitive to changes in the refractive index, thickness or crystallinity.

Differential interference contrast can be obtained by several methods, the ones most often used are Nomarski or Jamin-Lebedeff. The principle is that polarised light is passed through a Wollaston prism, where the light beams are slightly split, and after passing through the specimen the beams are recombined on the second prism. The image contrast is formed mainly on the boundaries where each beam has passed through regions of different optical properties. The boundaries appear dark or bright so that the object gives a three-dimensional impression. The resolution remains high and the sensitivity is comparable to phase contrast (see Figure 5.2b). The major disadvantage of this technique is cost and difficult alignment.

4. *Fluorescence.* Incident light microscopy can be utilised for fluorescence contrast. The incident light from a powerful and broad band source (mercury or xenon discharge lamp) is filtered to obtain a narrow band of an excitation wavelength. The emitted and reflected light is filtered again to allow only the longer fluorescence wavelength to form the image. This can be achieved by a set of filters and a semitransparent mirror mounted in a filter block. The emitted light is dependent on the fluorescence ability of the specimen. This technique is useful for the study of organic compounds as many conjugated structures fluoresce. Chemical staining, physical blending with a fluorescent dye, thermal degradation, oxidation or irradiation can be used to enhance the fluorescence of a particular phase. An example of an electron beam irradiated polymer blend is shown in Figure 5.2c. This technique has the advantage that the morphology is not influenced by the presence of a dye and that the phases can be identified.

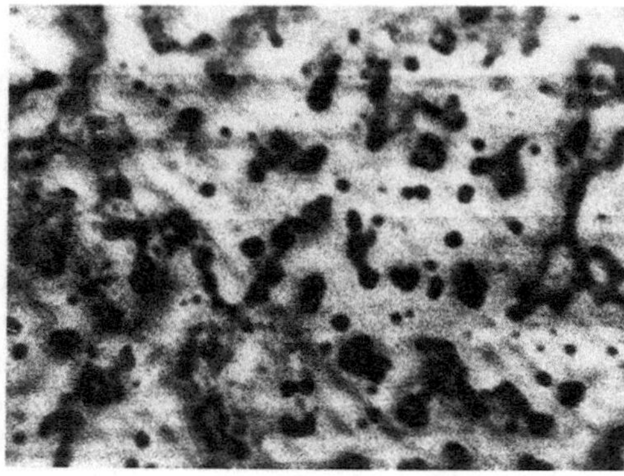

(a)

(b)

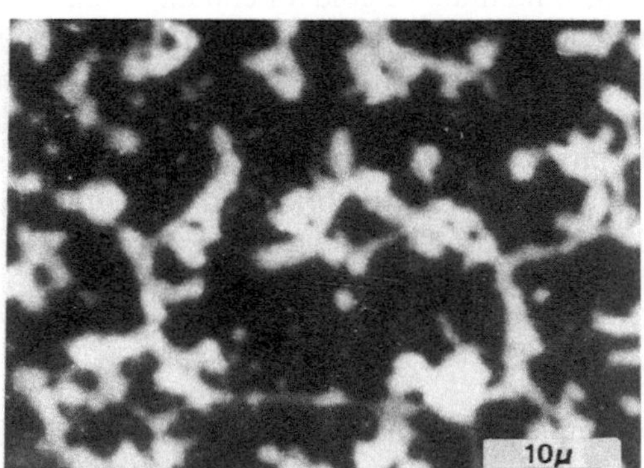

(c)

Figure 5.2 Light micrographs of PVC/PS blends using (a) phase contrast, (b) differential interference contrast and (c) fluorescence after electron beam irradiation.

5.1.3 Specimen preparation

Specimen preparation is the most important part of microscopy as the quality and the reliability of the results are dependent on the specimen. For transmission microscopy specimens must be flat, thin (the thickness must be less than the resolution required), clean and must have an optical contact with the cover slip (immersion oil is often used).

Specimens can be prepared by cutting on a microtome, by solvent casting or melt casting. Polymers are very soft materials and cutting or grinding often produces damaged surface layers, which obliterate the structure. It is therefore sometimes necessary to cut at low temperatures (− 40 °C or less), to press the sample onto a hot glass or to swell the surface with a suitable solvent. The stresses formed in a thin section as a consequence of the cutting often cause curling of the sample. The sample can be flattened again by heating it up using radiation from a hot source or placing the sample on a hot substrate (e.g. glycerine surface heated to about 60 or 80 °C. Hot pressing is also used to produce flat, scratch-free samples. Contrast on the scratch or cutting marks can be reduced by using an embedding oil or other medium of matching refractive index, making them less visible.

Contrast on polymer phases can be enhanced by staining (osmium tetroxide, lead citrate or uranyl acetate are used for polymers containing isolated double bonds). The double bonds can be formed in polymers by electron beam irradiation, by heat degradation or chemically by strong acids (chlorosulphonic acid).

For incident light microscopy the samples are prepared by cutting, mounting in a resin, surface grinding and polishing using a diamond paste. These samples are then etched to reveal certain morphology and vapour-coated with carbon, aluminium or gold for enhanced reflectivity. Etching can be done in strong acids (sulpho-phospho-chromic or chloro-sulphonic) or in solvents which do not swell the matrix. The latter technique often leaves a deposit which must be removed mechanically (gentle repolishing). Fracture surfaces can also reveal the phases but in some cases only when the phases are large, the interfacial bond weak and the matrix brittle. In most other cases the polymer structure is masked by the roughness of the fracture surface.

5.1.4 Applications

There are many examples in which light microscopy has been used for study of imiscible polymer blends; for example polyethylene/polypropylene blends were studied by polarised light microscopy [5, 6] and by fluorescence microscopy [4, 7, 8]. Excellent examples of interference contrast on polymer blends by Hoffman are in [4]. Surprisingly, very few publications cover the use of the best technique, which is the phase contrast [6]. This technique is useful for both amorphous and crystalline specimens and reveals the phases clearly (see

Figure 5.2a). It is obvious that the main difficulty remains the preparation of high quality undamaged specimens.

5.2 Scanning electron microscopy

The advantages of scanning electron microscopy (SEM) are a good depth of field, good resolution and easy specimen preparation. The cost of the instrument is also less than the cost of a transmission electron microscope. However, only information on surface topography can be obtained and contrast is usually not suitable for good image processing. These two disadvantages do not outweigh the advantages and the technique is very popular.

5.2.1 SEM instrument

The principle of a scanning electron microscope (Figure 5.3) is as follows. A small electron probe is formed by an electron gun and a set of three or more electromagnetic lenses. The electron gun consists of a heated tungsten filament, an anode attracting the emitted electrons and a focusing electrode (gun bias). The important parameter of the gun is its brightness which is given by the total number of electrons coming from a unit area of the source into a unit solid angle. In order to increase the brightness, it is possible to increase the

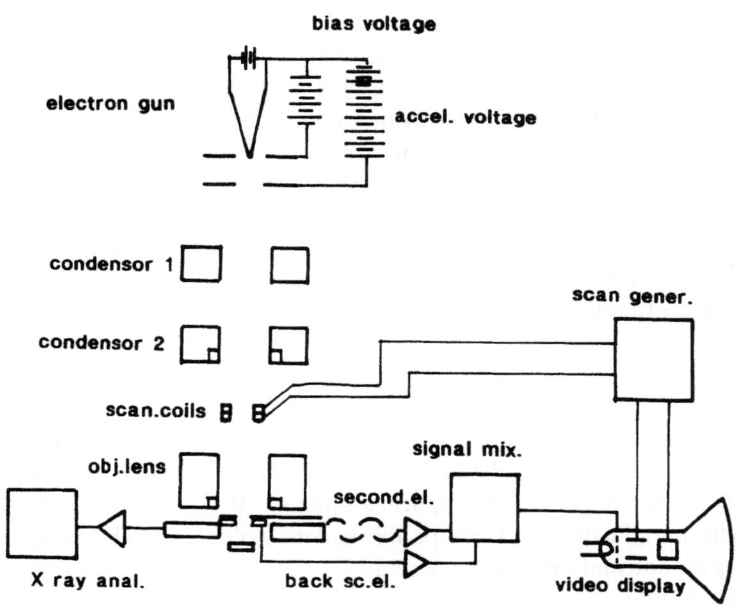

Figure 5.3 Schematic diagram of a scanning electron microscope.

emission current by replacing tungsten with lanthanum hexaborite. Alternatively, the area of the source can be reduced with a pointed filament or the emission angle can be reduced by field emission. The last source is the best for resolution, as a very small probe of less than a nanometer with a very high current can be formed. Unfortunately ultra-high vacuum is required for this source and the cost is very high.

The probe size is reduced by demagnification of the filament image using two electromagnetic lenses and then focusing onto the specimen surface by using a final (or objective) lens. The probe is scanned on the specimen by two sets of scanning coils controlled by the same scan generator as the cathode ray tube used to observe the image. The signal is detected by a low noise scintillator–photomultiplier–amplifier system and modulates the display signal. Each point of the scanned raster on the sample thus has a corresponding point on the display screen. Once the probe is focused and corrected for astigmatism, the magnification can be changed by changing the size of the scanned area without refocusing.

5.2.2 Specimen preparation

The microstructure of bulk specimens can be revealed by cutting or fracturing. Artefacts can be easily produced by both and a careful and critical assessment of the observed structure is needed. Fracture surfaces can reveal the microstructure only if the crack propagates through well-defined interfaces and the shape of the second phase is easily recognisable. Cut surfaces rarely reveal any microstructure, unless the mechanical properties are again very different, but often the beam damage (differential loss of mass) can reveal the second phase. Chemical etching or solvation is also used successfully. All specimens must be coated with a conductive layer of evaporated carbon or preferably sputtered gold. This coating will to some extent mask the structure of the polymer and will limit the resolution to about 10 nm. Magnification of more than 10 000 × is thus rarely used. Low accelerating voltages will reduce penetration of the conductive layer and thus minimise charging. Back-scattered electrons, which require high accelerating voltages and heavier elements, cannot thus be used for polymers to the same effect as for metals.

5.2.3 Application to polymer blends

The SEM can be used for immiscible polymer blends, where the phases are large, spherical and will debond easily. Examples include the work on PS/EPVM blends [9] or on toughened epoxies [10]. It is more difficult to obtain meaningful data from compatible and miscible blends where interfacial bonding is good and the phases are small. In this case it is difficult to recognise the phase boundaries and eliminate damage effects, such as crazing, which will produce structure of their own and which can be mistaken for a second phase.

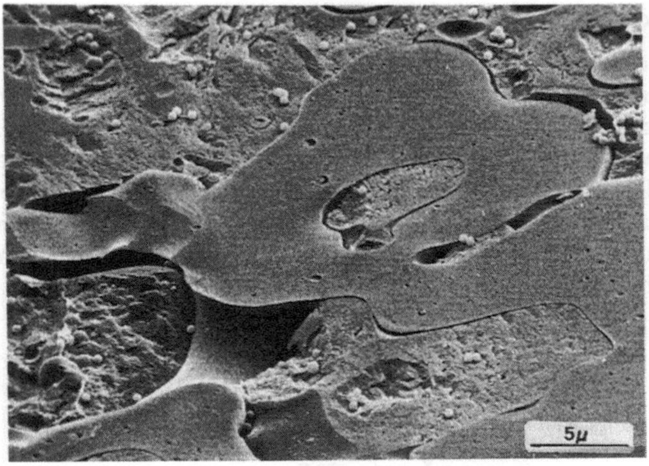

Figure 5.4 SEM micrograph of PE/PC blend (S.Woodisse and D. Vesely).

Cutting and etching is a more reliable sample preparation method as it can differentiate between different polymeric materials. The resolution is about 20 nm, which is 50 times better than light microscopy, however the conductive coating layer will reduce the resolution further. A good example is shown in Figure 5.4. A very useful overview of SEM studies of polymer surfaces is contained in [11].

5.3 Transmission electron microscopy

Transmission electron microscopy (TEM) represents the ultimate technique in terms of resolution and analysis of materials. It is the only technique which provides information on the fine structure of materials down to atomic or molecular levels and elemental analysis from small volumes. In comparison with other imaging techniques TEM has some advantages (Table 5.1).

Table 5.1 Advantages of transmission electron microscopy

	Resolution for polymers	Cost (relative) (LM = £25k)	Specimen preparation	Analysis
LM	1000 nm	1	Easy/difficult	Qualitative
SEM	20 nm	6	Easy	EDX, WDX
TEM	1 nm	12	Very difficult	EDX, EELS, contrast diffraction

5.3.1 The microscope

The microscope (Figure 5.5) is similar in principle to the SEM, i.e. it is based on the electron gun, which provides a narrow beam of electrons with good brightness and accelerating voltages of generally 50 to 100 keV, while 200, 300 and even 400 keV microscopes are now also commercially available. The current density of the electron beam on the sample can be controlled by gun bias, condenser lens aperture and defocus of the condenser lenses. The specimen stage is designed to take sample support grids 3 mm in diameter and 0.1 mm thick and provides tilt and x–y shift of the sample, and even rotation.

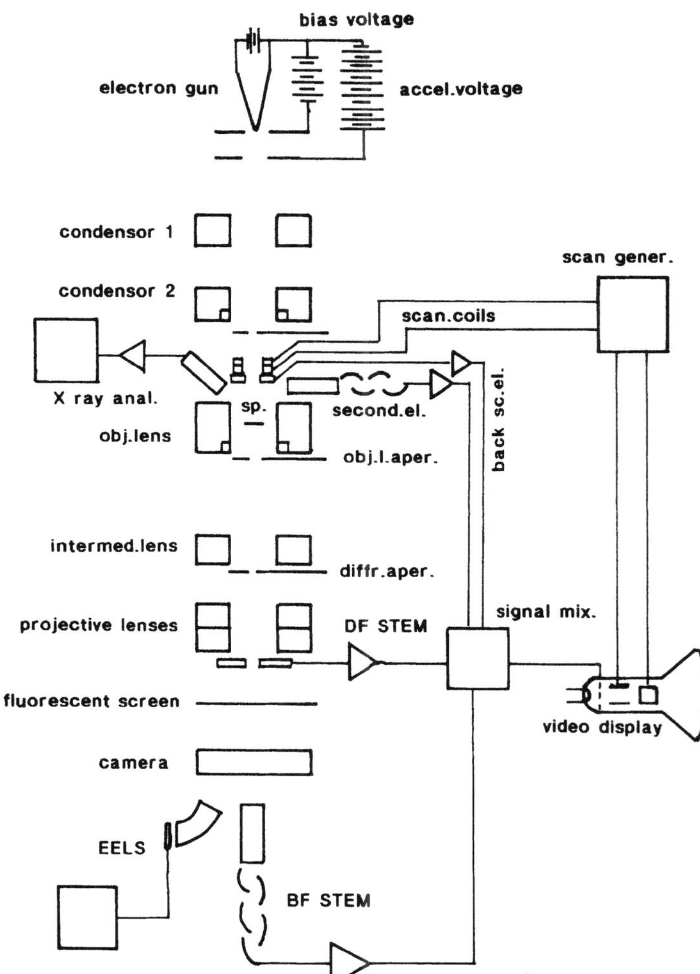

Figure 5.5 Schematic diagram of a transmission electron microscope with scanning and analytical facilities.

Side entry stages are most popular, although less stable than the top entry ones, as they are easier to use and can be equipped with auxiliary functions (cooling, heating or straining stages, environmental cell). The specimen chamber is provided with cryopumping and/or an ion pump. The stage is placed in the centre of the objective lens which is used for focusing. The magnification of up to 300 000 × is obtained by two projective lenses. The intermediate lens also provides some magnification but is mainly used to change the focusing point from the specimen (image) to the back focal plane of the objective lens (diffraction). The objective lens aperture is used to cut off the scattered or diffracted beams and thus enhance contrast. The aperture in the intermediate lens is used to select the diffraction area. The image is viewed on a fluorescent screen which converts electron energy to visible light. the photographic camera is placed under the fluorescent screen and is basically a storage and transport mechanism for the photographic plates. The alignment and astigmatism correction on modern microscopes is now done by electromagnetic deflectors. The auxiliary equipment needed consists of a vacuum pumping system (vacuum 10^{-5} torr) and the power supply (stability 10^{-6} is required).

5.3.2 Specimen preparation

Unlike metals, polymers cannot be prepared by the chemical or the electrochemical method, as they cannot be dissolved homogeneously without swelling. Cutting is the only technique which can be used for preparation of thin sections from bulk polymers. Cutting is performed using an ultramicrotome with a glass or diamond knife (Figure 5.6). The knife has a water trough for collecting thin sections and the sample is moved up and down with a small advance for each cut. The sections are picked up from the water surface on a copper supporting grid. There is always some distortion of the sections by compression and the surface tension of water is not always capable of stretching them out again. Crystalline polymers are the most difficult to cut and hardening of the polymer by cooling to low temperatures or chemical modification might be required [12].

It is advisable to vapour-coat the polymeric samples with a thin layer of carbon to reduce charging. This will not only support the film but it will also conduct away the charge and heat generated by the electron beam. It will to some extent reduce the rate of mass loss [13]. Vacuum evaporation of carbon is, however, efficient only above 3000 °C and the specimen, which is only a few centimetres away, will be exposed to an extreme heat. This often results in severe heat damage of a polymeric sample and in order to minimise this effect a special technique has been developed [14, 15] which utilises a carbon fibre thread and an electronic shutter. Temperature measurements have shown that the specimen will not be heated above 40 °C when this technique is used and it also has the advantage that the thickness of the carbon layer is very reproducible.

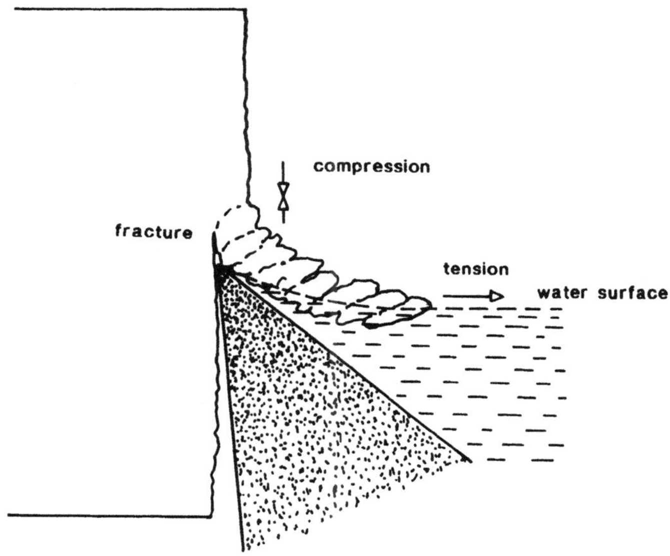

Figure 5.6 Illustration of major problems with microtomy of polymers.

Solvent casting and melt casting are also very useful techniques. The polymer is dissolved in a suitable solvent and cast on a glass slide. After a slow, controlled evaporation, the film is removed onto a water surface and mounted on a copper grid. It is important to remember that the structure is different from that of the bulk material and that artefacts can be created.

Replication of a surface is a simple and very rewarding technique. It provides information on the surface morphology identical to that obtained using scanning electron microscopy but with better resolution. The two-stage replica is prepared by taking the imprint of the surface using a softened material, e.g. gelatin, polyacrylic acid or acetal, coating the surface with carbon and then evaporating a heavy metal at a small angle (15–20 °C) onto it. The substrate is then dissolved and the replica is picked up on to a copper supporting grid.

Staining of samples can be performed in solution or in vapour. The vapour phase staining has the advantage that the sample cannot be so easily damaged or contaminated as in liquids. For polymers containing isolated double bonds osmium tetroxide is the most common stain [16, 17]. In order to reduce the health risk associated with this stain, a special staining vessel has been constructed [18]. The use of this vessel not only makes the handling of osmium safer but it also substantially reduces the consumption of this expensive chemical.

5.3.3 Contrast

Contrast is used not only to visualise the structure but also to obtain more information about the specimen. Careful study of contrast effects is thus of significant importance.

(a) Mass thickness contrast is dependent mainly on the scattering power of the sample and is most important for microscopy of amorphous materials. The objective lens aperture is used to eliminate the scattered electrons from the image formation. The contrast is thus produced by the difference between the intensity of the scattered and the unscattered beams. The scattering power of a particular area on the sample is directly related to the number of atoms per electron path (specimen thickness) and their scattering power (atomic number). The heavy elements thus provide higher contrast. This effect is utilised in replica and in staining contrast (see Figure 5.7).

(b) Diffraction contrast is much stronger as the diffracted intensity can account for more than 50% of the total. The objective lens aperture is again used to eliminate the diffracted beam. As the diffraction is critically dependent on the orientation and the periodicity of the interatomic planes, all imperfections will appear in strong contrast for suitable orientations and their crystallographic geometry can be studied. There are several contrast effects which are observed.

With increasing thickness more and more electrons are diffracted, until the total intensity is diffracted outside the objective lens aperture and the area becomes dark. For further increases in the thickness the electrons can be diffracted twice and this area brightens again. This

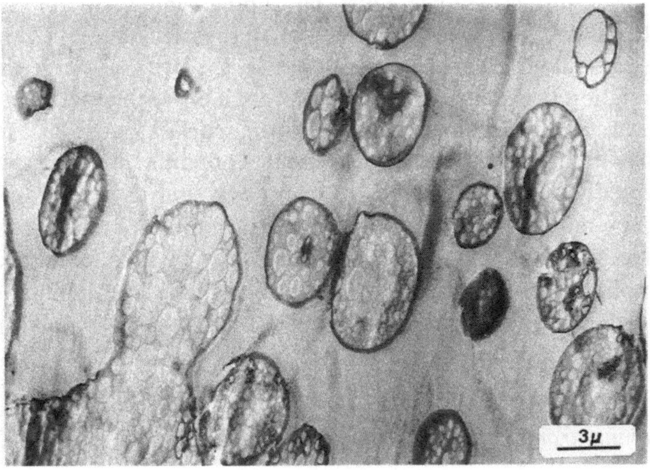

Figure 5.7 TEM micrograph of cut and OsO_4 stained ABS sample.

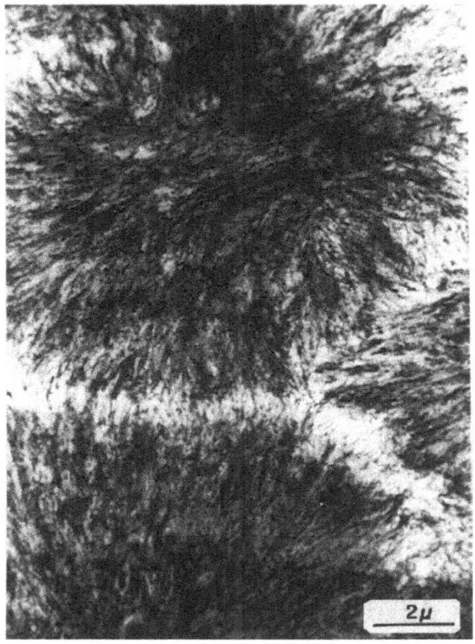

Figure 5.8 STEM micrograph of solvent cast specimen of PE/PS blend. The mass–thickness contrast is much weaker than diffraction contrast.

means that only areas of the specimens of a certain thickness are transparent for given diffraction conditions. A similar effect can be observed for specimens which are bent (bending fringes). Only regions which are oriented for the Bragg condition appear dark.

The contrast studies must be supplemented by the diffraction patterns from the investigated region of the specimen. These will reveal the crystallographic structure and also the orientation of the sample in relation to the incident electron beam. Diffraction patterns also help to select the individual diffracted beams for imaging (dark field).

Crystallographic studies on polymers are difficult as the crystallinity is rapidly destroyed by the electron beam and often specialised techniques must be used. The complexity of the contrast in crystalline polymers is shown in Figure 5.8 for a solvent cast specimen. The cut samples cannot be used for any serious crystallographic studies and for this reason electron microscopy has been done mainly on solvent or melt cost samples.

5.3.4 Scanning transmission electron microscopy (STEM)

The scanning facilities can be added to a modern transmission microscope without difficulty. A secondary electron detector is placed above the objective lens and the electrons must climb out of the lens to reach it. This will reduce to some extent the signal but the resolution obtained is superior to conventional SEM (1.5 nm compared to 10 nm) as the specimen is inside the lens. In addition, transmitted electrons can also be detected and this provides a very valuable technique, which has the following advantages:

- *Image processing.* As the image is constructed point by point from the signal, electronic image processing and contrast enhancement is possible. This is very useful for weak contrast as obtained for polymer blends (see Figure 5.9).
- *Beam damage.* Another advantage is the reduction in the beam damage as low beam currents can be used to form the image (image intensification). The spot scanning technique with high resolution is essential for minimising the effect of specimen movements.
- *Microdiffraction.* The highly excited objective lens and the position of the sample enable a parallel beam of small diameter to be obtained. This will provide a very accurate, high quality diffraction pattern, which can be precisely located on the area of interest.

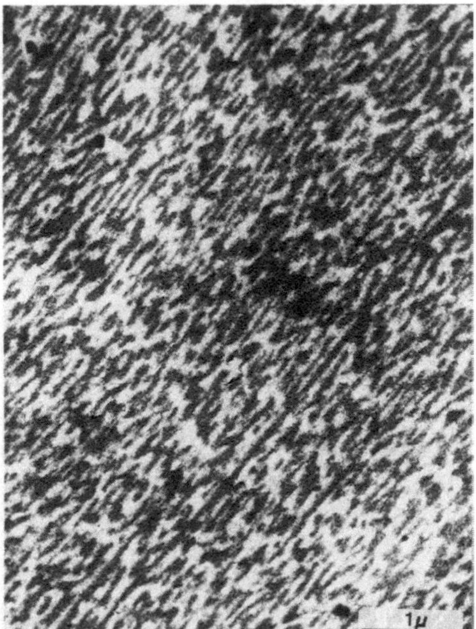

Figure 5.9 STEM micrograph of cut sample of PVC/SAN blend. Contrast has been enhanced by signal mixing.

- *Other signals* can also be utilised, such as back-scattered electrons, forward-scattered or diffracted electrons (annular detector), and signals can be combined to increase contrast.
- *Convergent beam pattern.* The STEM configuration can also be used to obtain a highly convergent beam focused on a very small area (about 2 nm). This will result in a pattern which can be used for very accurate crystallographic studies. However, the required exposures are high for most beam sensitive polymers.

5.3.5 Electron beam damage

Polymers in the electron microscope undergo rapid and substantial chemical changes as a result of interaction with a highly ionising radiation of the electron beam. Polymer chains are decomposing to more stable structures, often conjugated, and also form volatile products. This results in two major effects: loss of mass and formation of double bonds. It has been shown [19–22] that the decomposition is a first order reaction and can therefore be expressed as an exponential or a sum of exponential curves [23].

The mass loss can be measured by the decrease in the scattering power of the sample or the increase in the transmitted intensity. A Faraday cage below the specimen will provide a sufficient accuracy when connected to a sensitive current meter. A narrow beam, selected by a small objective lens aperture, will provide a sufficient contrast between the progressive stages of irradiation (resolution). The loss of mass with exposure can be calculated from the transmitted intensity taking into account the thickness of the carbon coating layer [24]. It is possible to measure the mass loss from an area as small as 2 μm in diameter using the STEM microdiffraction mode. This technique can be utilised for identification of phases as different polymers lose mass at a different rate and have a different terminal residue [25]. It has been shown previously [26] that this technique can also provide some information on partial miscibility in polymer blends. An example of the mass loss measurement is shown in Figure 5.10 for PVC/SAN blend.

The electron beam damage can also be utilised for contrast enhancement and identification of phases on a very fine scale (phases of about 20 nm can be resolved) [27–29]. As part of the polymer degradation process isolated double bonds are formed on the polymer chain backbone. These bonds can be stained with osmium tetroxide [16] and their density is dependent on the chemical nature of the polymer. With progressing irradiation the isolated double bonds are joined to form conjugated sequences and the staining intensity is reduced. This means that the staining intensity will reach a maximum at a particular exposure, which is a characteristic feature of a given polymer compound. It is possible to select an exposure at which the phases have a known and distinct contrast and can thus be positively identified.

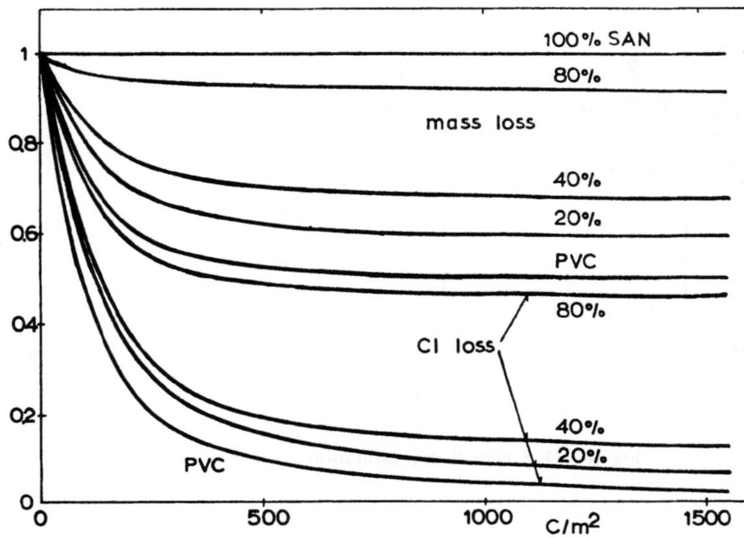

Figure 5.10 Mass loss from PVC/SAN blends.

5.3.6 Applications to polymer blends

Transmission electron microscopy can be used to obtain structural information on a small scale approaching the molecular dimensions of the material. A good example is the study of the microstructure of block copolymers [30]. It can be used to study surface morphology as well as internal structure and can provide crystallographic information. Crystallographic information must be obtained before the specimen is damaged and this requires special techniques such as computer control of the exposure, of the microscope alignment (rapid change of the mode of operation) and of the beam position. Image intensification, frame store and subsequent image processing are very useful additional techniques. These techniques can provide information on co-crystallinity, very fine dispersion of phases [21] and on interfaces [31]. For amorphous materials or the study of phases in polymer blends, specimens can be stained, irradiated or etched and stained or the contrast can be enhanced electronically using STEM techniques. This will unfortunately limit the resolution to about 3 nm or more. The preparation of thin specimens of high quality is the major problem in transmission electron microscopy and requires a high degree of knowledge, experience and patience.

5.4 Thermal analysis

Several structural transitions can occur in polymers during heating. These transitions are, for example, melting and crystallisation, glass transition temperature (associated with the change in the specific heat capacity and thus

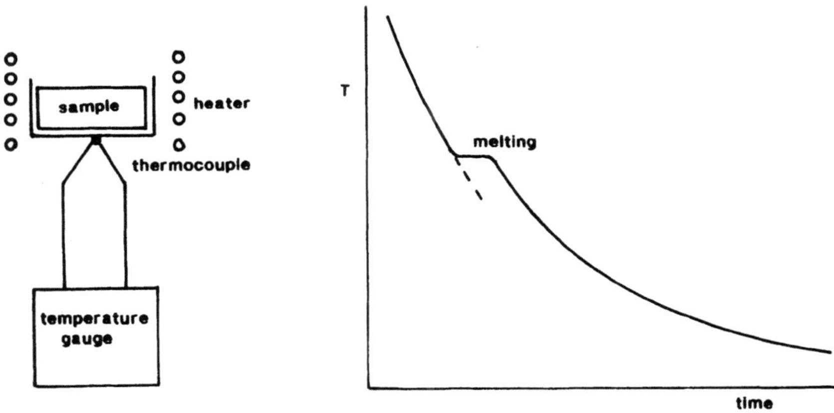

Figure 5.11 Schematic representation of thermal analysis.

with the freedom of molecular motions), crystal transformations, decomposition and other chemical reactions, volatilisation and annealing effects. For characterisation of polymer blends the most important transitions are the glass transition temperature and the melting point. Polymer mixtures and also block copolymers retain the properties of the individual homopolymers, but random copolymers and miscible polymers behave as a single compound. It is thus possible to obtain some information on the degree of mixing, copolymerisation, co-crystallisation and other interactions on a molecular level. The principle of these measurements is illustrated in Figure 5.11. For a constant rate of cooling the changes in the temperature of an empty cell will follow an exponential curve. When a sample is placed in the cell, its thermal capacity will slow down the rate of cooling as more heat has to be extracted, and a different exponential curve will be followed. The difference between these two rates of cooling is related to the mass and the specific heat of the sample. Any change in the specific heat or any extra absorption or evolution of heat will be registered as a deviation from this difference.

5.4.1 Differential thermal analysis (DTA)

It is more convenient to measure the differences in the cooling (or heating) curves simultaneously by using two identical cells, one for the reference curve and one for the sample curve. The basic principle of this technique is illustrated in Figure 5.12. Only one heater is needed and both the reference and the sample cells are heated simultaneously. The sample will add or remove heat from its environment and the temperature in the sample cell with thus be different from the temperature in the reference cell. The difference will to some extent be dependent on the rate of cooling or heating. The computer

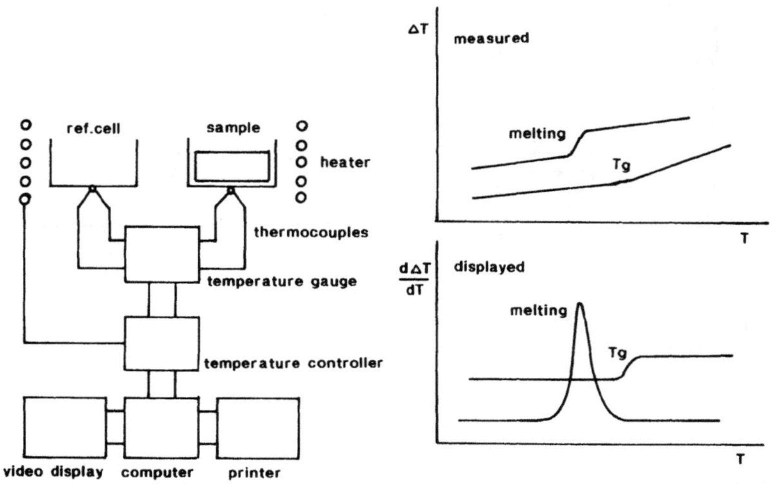

Figure 5.12 Schematic diagram of DTA.

analysis of the data can provide information very similar to that obtained by the more sophisticated differential scanning calorimetry.

5.4.2 *Differential scanning calorimeter (DSC)*

Structural changes are usually associated with changes in heat absorption or emission and are measured using calorimetry. Scanning calorimetry can also measure changes at constant heating or cooling rates. The differences in heat loss or gain between the sample and the reference cells are measured in a differential scanning calorimeter. In this case the heat input needed to maintain both cells at the same temperature is measured. The principle of this technique is schematically illustrated in Figure 5.13. The cells are mounted in a metal block, which can be cooled for more efficient heat stability. The thermocouples measure the temperature as well as differences in the temperature. The difference is quickly compensated by the heat controller and the amount of compensation is measured. From these measurements the rate of heat absorbed by or evolved from the sample can be computed.

5.4.3 *Glass transition temperature*

The sudden change in specific heat capacity at T_g is manifested as a step on the temperature versus differential heat absorbed (or differential temperature) curve (see Figure 5.10). The material usually has different mechanical and other properties on both sides of this transition. The temperature at which this transition occurs is different for different materials. In physical blends of two polymers each polymer retains its own transition temperature and both are

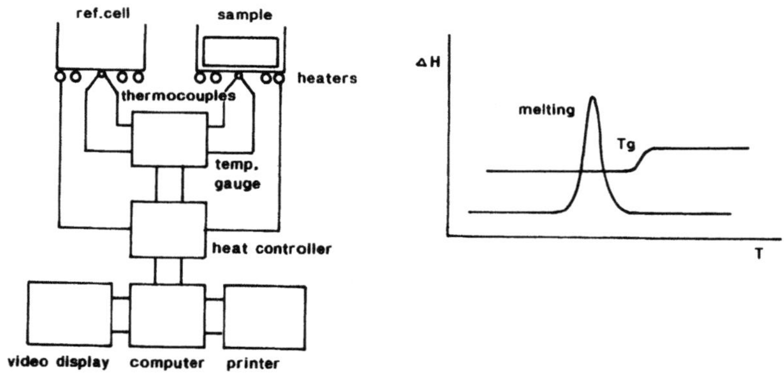

Figure 5.13 Schematic diagram of DSC.

observed, providing that they are further apart than the resolution of the analytical technique used. The presence of one glass transition temperature for a blend is an indication of this homogeneity on a molecular level and thus mechanical integrity. The measurement of glass transition temperature can therefore assist in determination of compatibility of amorphous polymer blends.

5.4.4 Practical aspects

The accuracy of these techniques is dependent on the proximity of the T_g values (which should be more than 20 °C), on the sharpness of these transitions and on the sensitivity, accuracy and resolution of the experimental technique used. For small concentrations (less than 10%) the small peak is difficult to resolve and the shift of the major peak is negligible. It is essential that the sample has a good and constant contact with the container so that the heat exchange is as good as possible. Preforming and premelting of the sample is often necessary. This also removes the built-in stresses due to cutting, forming and mixing, as structural relaxation occurs at T_g and can distort the shape of the transition. The thermal history of the sample must therefore be known and its effect on the transition region understood. The effect of additives on the T_g can also be significant. Blending with another polymer can result in the redistribution of additives, their segregation or reabsorption by the second polymer.

In conclusion, the thermal analysis techniques are very useful in some well-defined cases where the glass transition temperatures are well separated, sharp and sufficiently large. Unfortunately such systems are not often miscible. Also the use of these techniques for systems where one or both components are crystalline is not straightforward. Combination of techniques is always a necessity.

5.5 Other techniques

5.5.1 Light scattering

Light scattering is a very simple technique which in principle consists of a good light source (laser) and a sensitive detector (Figure 5.14). The detector can be fixed at 45 or 90° or be movable through a range of angles. The light entering the sample cell will be scattered on heterogeneities such as large molecules or molecular clusters. The transmitted intensity will thus be reduced and the scattered intensity increased. This technique can detect segregation of phases in solutions with temperature changes (e.g. spinodal decomposition, cloud point), but can also be used for detection and even measurement of phase size and dispersion in thin films. The simplicity of this technique is an attraction but the interpretation of the data is not always straightforward.

5.5.2 Neutron and X-ray scattering

Scattering techniques with shorter wavelengths require deuterated polymers for neutrons and at least fluorinated polymers for X-ray scattering. These techniques are therefore rather specialised and are not commonly used for polymer blends. The experimental set-up is similar to that for light scattering. The information obtained is on the molecular scale and can reveal the arrangement of side groups, crystalline structure and conformation of molecules. It can be related to dispersion of phases but a reliable interpretation of data is not always possible.

5.5.3 Spectroscopy

The spectroscopic techniques based on molecular vibrations can measure molecular interactions, such as hydrogen bonds (FTIR, Raman, NMR) or

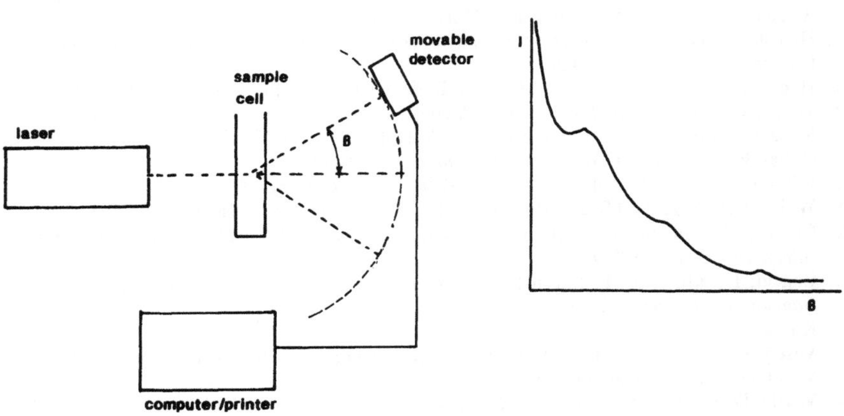

Figure 5.14 The principle of light scattering.

chemical reactions. However, these techniques are not very sensitive to the phase dispersion and are mostly non-quantitative. Their use for the study of polymer blends is therefore limited and they must be used in conjunction with other techniques.

It is often important to gain some information on interfacial interactions but it is necessary to differentiate between interaction at interfaces and in the bulk due to partial miscibility. The location of the signal can only be obtained with probe techniques (e.g. FTIR-microscopy) which have a spatial resolution of 5 μm or more.

5.6 Conclusions

The above description of techniques used for study of polymer blends deals only with those techniques which are most direct and easy to use. There are, however, many other techniques which are specialised and which require expert interpretation. Practically all available techniques have been tried or used for investigation of polymer blends with some more or less useful results.

The selection of a suitable technique is dependent on the information desired. The most common information required is the detailed knowledge of phase dispersion for which the microscopic techniques are essential. The information on interfacial interaction and on partial solubility can be best obtained with FTIR or Raman microscopy. However, it is always useful to compare the results of several techniques with the properties of the bulk material.

References

1. Bucknall, C.B. (1977) *Toughened Plastics*, Applied Science Publishers, London.
2. Shaw, M.T. (1985) in *Polymer Blends and Mixtures* Walsh, D., Higgins, J. and Maconnachie A., eds., Martinus Nijhoff Publishers, Dordrecht, p. 37.
3. Hemsley, D. (1984) *Light Microscopy of Synthetic Polymers, RMS Microscopy Handbooks 07*, Oxford University Press, Oxford.
4. Hemsley, D. (ed.) (1989) *Applied Polymer Light Microscopy*, Elsevier Science Publishers.
5. Teh, H.J. (1983) *J. Appl. Polymer Sci.* **28**, 605.
6. Korger-Kocsis, J., Kalo, A. and Kuleznev, V. (1984) *Polymer* **25**, 279.
7. Gelles, R. and Frank, C.W. (1983) *Macromolecules* **16**, 1448.
8. Billingham, N.C. and Calvert, P.,D. (1989) *Inst. Phys. Conf. Ser.* **98**, 571.
9. Walsh, D.J., Higgins, J.S. and Rostami, S. (1983) *Macromolecules* **16**, 388.
10. Chan, L.C., Gillham, J.K., Kinloch, A.J. and Shaw, S.J. (1984) Rubber toughened epoxies. *Advanced Chem., Ser.* **209**.
11. Engeld, L., Klingele, H., Ehrenstein, G.W. and Schopen, H. (1981) *An Atlas of Polymer Damage*, Wolfe Science Books.
12. Kanig, G., (1974) *Kunststoffe* **64**, 470.
13. Vesely, D., Low, A. and Bevis, M.J. (1975) *Developments in Electron Microscopy and Analysis*, Venables, J.A., ed., Academic Press.
14. Vesely, D. and Woodisse, S. (1982) *Proceedings RMS* **17**, 137.
15. Finch, D.S. and Vesely, D. (1985) *Inst. Phys. Conf. Ser.* **78**, 151.
16. Andrews, E.H. and Stubbs, J.M. (1964) *J. Roy. Micr. Soc.* **82**, 221.

17. Kato, K. (1965) *J. Electron Micros.* **14**, 220.
18. Owen, G. and Vesely, D. (1985) *Proceedings RMS* **20**, 297.
19. Vesely, D. and Lindberg, K.A.H. (1983) *Inst. Phys. Conf. Ser.* **68**, 7.
20. Vesely, D. (1984) *Ultramicroscopy* **14**, 279.
21. Vesely, D. and Finch, D.S. (1988) *Macromol. Chem. Symp.* **16**, 329.
22. Finch, D.S. and Vesely, D. (1987) *Polymer* **28**, 329.
23. Vesely, D. (1990) *Proc. XII Int. Congress on EM, Seattle*, San Francisco Press **2**, 410.
24. Vesely, D. (1988) *Inst. Phys. Conf. Ser.* **93**, 455.
25. Lindberg, K.A.H., Vesely, D. and Bertilsson, H.E. (1989) *J. Mater. Sci.* **24**, 2825.
26. Vesely, D. and Finch, D.S. (1987) *Ultramicroscopy* **23**, 329.
27. Parker, M.A. and Vesely, D. (1983) *Inst. Phys. Conf. Ser.* **68**, 11.
28. Vesely, D. (1984) *Ultramicroscopy* **14**, 279.
29. Parker, M.A. and Vesely, D. (1992) *J. Elec. Micr. Techniques* (to be published).
30. Hadlin, D.L. and Thomas, E.L. (1983) *Macromolecules* **16**, 1514.
31. Vesely, D. and Ronca, G. (1985) *Inst. Phys. Conf. Ser.* **78**, 423.

6 Theoretical aspects of polymer blends and alloys

R.G.C. ARRIDGE

An attempt has been made in this chapter to provide a sound basis for understanding the mechanical behaviour of polymer blends and alloys. The presentation has not been over-simplified, for the author agrees with the point of view of Treloar [1] in the preface to the first edition of his book *The Physics of Rubber Elasticity* that '... if one is going to have a theory at all one may as well take some trouble to find the one which most nearly represents the known facts.' It is relatively easy to produce equations which *describe* the behaviour of blends and alloys (and, indeed, of polymers in general) sufficiently well for the operation of, for example, extrusion machines or for the interpretation of standard test procedures on the solid material. However, it is not at all easy to *explain* their behaviour in a rational and scientific manner such that prediction of properties is possible and the understanding of materials scientists and engineers is increased. This the author has attempted to do in this short chapter but, for the serious materials scientist or technologist, the chapter should in any case be looked upon as a guide to further study, to the literature (references to some of which are given) and to the ever-increasing amount of new knowledge that is being gathered about these interesting and useful materials. Only in this way can progress be made to develop newer materials with novel properties.

6.1 Introduction

This chapter describes the elastic and viscoelastic properties of polymer blends and alloys, and some definitions are relevant at this stage. *Elastic* refers to deformations under load which are reversible, however large they may be (as in rubbery materials, for example). *Viscoelastic* applies to reversible deformations which are time-dependent, that is, that not only the deformation but also its recovery may take an appreciable time to complete. Such properties are typical of all polymers, they are also temperature sensitive and it is usual to consider them as linear, that is the deformation under any load scales linearly with the load. *Anelastic* is used by some authors to define *linear recoverable* viscoelasticity as a sub-group of a more general viscoelasticity. Certain results and definitions from the theory of elasticity will be required and are given here for convenience. For more details one of the references listed may be consulted (e.g. [2, 3]).

In classical elasticity the load per unit area of the undeformed material is termed *stress* and the deformation, divided by the original length, is a dimensionless ratio termed *strain*. Hooke's law relates stress to strain by a *modulus*. For simple extension this is Young's modulus.

For more complex types of loading stress and strain need to be defined more generally. To do this correctly it is best to use tensor notation, defining stress and strain as second-rank tensors and the general relation between them as a fourth-rank tensor. For a comprehensive account of this the reader is referred to references [4, 5]. Here the results are summarised in a simpler notation.

First, however, it can be seen that both stress and strain in a three-dimensional body are neither scalar nor vector quantities. Stress is defined as a load divided by an area and load is a vector quantity with three components. The area on which it acts may also be characterised by a vector (in the direction of its normal and of magnitude equal to the area) but we cannot divide one vector by another. The quantity that is used to describe a stress is a second rank *tensor*, with nine components in general (three for load and three for area) which reduce to six in practice because of the equilibrium of forces. In the literature, stress is therefore written as σ_{ij}, where the first suffix refers to the direction of the normal to the area, the second to the direction of the force. Thus σ_{13} is a force in the 3 direction acting on a plane of unit area perpendicular to the 1 direction, σ_{22} a force in the 2 direction acting on a plane of unit area perpendicular to the 2 direction (Figure 6.1).

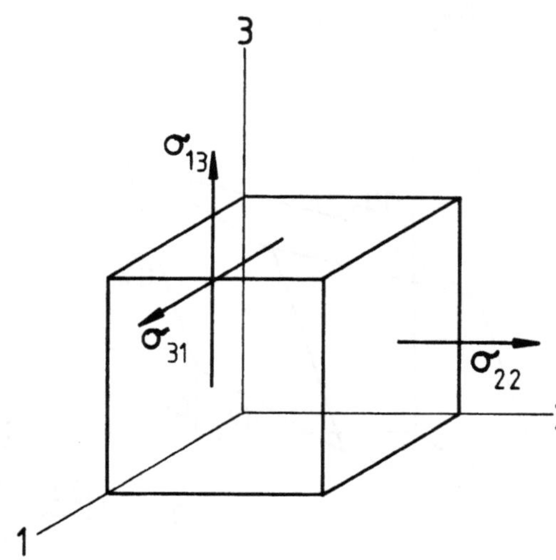

Figure 6.1 The description of stress in a solid.

An abbreviated notation is frequently used:

$$\sigma_1 = \sigma_{11}, \quad \sigma_2 = \sigma_{22}, \quad \sigma_3 = \sigma_{33}$$

and

$$\sigma_4 = \sigma_{23}, \quad \sigma_5 = \sigma_{31}, \quad \sigma_6 = \sigma_{12}$$

For strains two suffixes are also needed, one for the direction of the deformation, the other for the direction of the original line element. Figure 6.2 illustrates this for e_{22} (tensile) and e_{23} (shear) strains. There are nine components of strain but, again, these reduce to six because of symmetry.

The abbreviated notation for strain is not quite as simple as that for stress because of historical definitions. We have:

$$e_1 = e_{11}, \quad e_2 = e_{22}, \quad e_3 = e_{33}$$

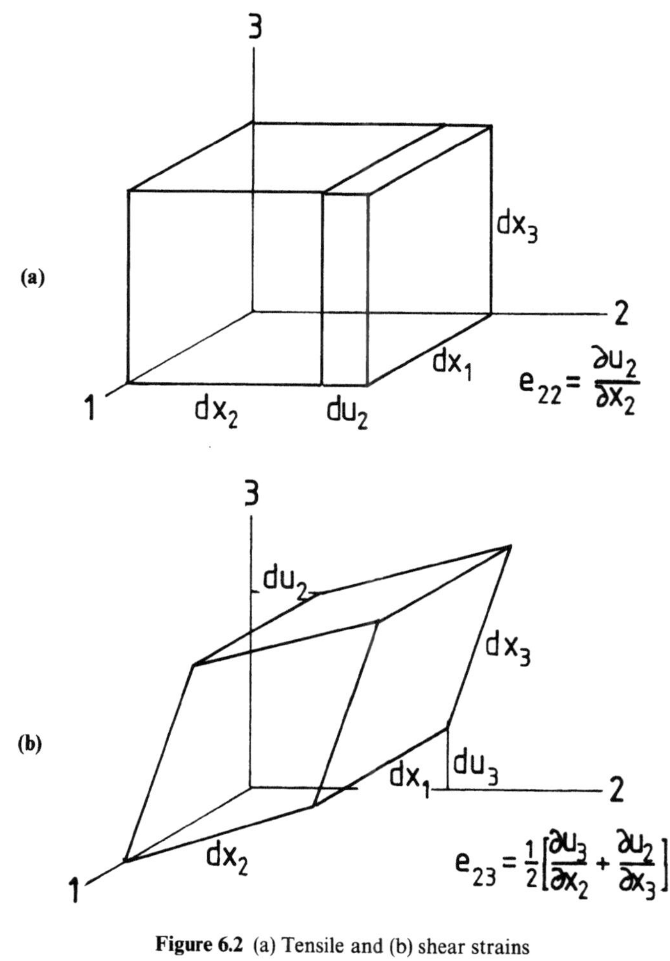

Figure 6.2 (a) Tensile and (b) shear strains

but

$$e_4 = 2e_{23}, \quad e_5 = 2e_{31}, \quad e_6 = 2e_{12}.$$

Now consider a uniaxial stretch in a simple isotropic material, that is, one in which the properties are the same in all directions. Thus stress σ_1 produces strain

$$e_1 = \sigma_1/E$$

where E is Young's modulus, but it also causes Poisson contractions $-\nu e_1$, in both 2 and 3 directions. Put another way we can write

$$e_1 = \frac{1}{E}[\sigma_1 - \nu(\sigma_2 + \sigma_3)]$$

since stresses σ_2 and σ_3 would produce strains along the 1 direction of

$$\frac{-\nu\sigma_2}{E} \quad \text{and} \quad \frac{-\nu\sigma_3}{E}$$

respectively.

Similar equations hold for e_2 and e_3. For shears we have

$$\sigma_i = Ge_i$$

where G is the shear modulus and i can be 4, 5 or 6. Now sum the tensile strains. *Dilatation* is written as:

$$\Delta = e_1 + e_2 + e_3 = \frac{(1 - 2\nu)}{E}(\sigma_1 + \sigma_2 + \sigma_3)$$

or

$$\Sigma\sigma = \sigma_1 + \sigma_2 + \sigma_3 = \frac{E\Delta}{1 - 2\nu}$$

We can then write

$$e_1 = \frac{1}{E}[\sigma_1(1 + \nu) - \nu\Sigma\sigma]$$

or

$$\sigma_1 = \frac{E}{1 + \nu}e_1 + \frac{\nu E\Delta}{(1 + \nu)(1 - 2\nu)}.$$

The term

$$\frac{\nu E}{(1 + \nu)(1 - 2\nu)}$$

is called Lamé's constant and usually written λ. Also

$$\frac{E}{1 + \nu} = 2G$$

where G is the shear modulus. So we can summarise:

$$\sigma_i = 2Ge_i + \lambda\Delta \qquad i = 1, 2 \text{ or } 3$$
$$\sigma_i = Ge_i \qquad i = 4, 5 \text{ or } 6.$$

(Note, in the tensor notation, these two equations combine to give $\sigma_i = 2Ge_{ij} + \lambda\Delta\delta_{ij}$, where δ_{ij} is Kronecker's δ defined as follows: $\delta_{ij} = 1$, $i = j$; $\delta_{ij} = 0$, $i \neq j$).

If, as in a hydrostatic deformation, $\sigma_1 = \sigma_2 = \sigma_3 = -p$ then we have

$$\Sigma\sigma = -3p = \frac{E\Delta}{1 - 2\nu} \quad \text{or} \quad p = \frac{-E}{3(1 - 2\nu)}\Delta.$$

This shows that the bulk deformation $\Delta = e_1 + e_2 + e_3$ (the dilatation) is proportional to the pressure p, with the factor of proportionality being the *bulk modulus*

$$K = \frac{E}{3(1 - 2\nu)}$$

(The negative sign occurs here because pressure causes *contraction* of the volume.)

For isotropic materials this modulus and the shear modulus G are the two important, independent quantities from which other moduli can be derived. Thus

$$E = \frac{9KG}{3K + G} \quad \text{and} \quad \nu = \frac{3K - 2G}{6K + 2G}$$

In rubbers and in other amorphous polymers above T_g, $K \gg G$ and the material can be considered quasi-incompressible.

In an *anisotropic* material the generalised Hooke's law has to include the relation between stress σ_1, say, and *all* the strains $e_1, e_2 \ldots e_6$. So, for example

$$\sigma_1 = C_{11}e_1 + C_{12}e_2 + C_{13}e_3 + C_{14}e_4 + C_{15}e_5 + C_{16}e_6$$

where C_{11}, C_{12}, etc. are constants, the elastic moduli. We may conveniently write the relation between all the stresses and strains, using matrix notation as

$$\sigma = \mathbf{C}\mathbf{e} \quad \text{or} \quad \mathbf{e} = \mathbf{S}\sigma$$

where $\mathbf{S} = \mathbf{C}^{-1}$. In these equations $\sigma = (\sigma_1\sigma_2\sigma_3\sigma_4\sigma_5\sigma_6)^T$ with an analogous vector $\mathbf{e}$ and $\mathbf{C}$ is the matrix

$$\begin{bmatrix} C_{11} & C_{12} & C_{13} & C_{14} & C_{15} & C_{16} \\ C_{21} & C_{22} & C_{23} & \cdot & \cdot & \cdot \\ C_{31} & \cdot & \cdot & \cdot & \cdot & \cdot \\ \cdot & & & & & \\ \cdot & & & & & \\ \cdot & & & & & \\ C_{61} & \cdot & \cdot & \cdot & \cdot & \cdot \end{bmatrix}$$

S is the inverse of this matrix and the S_{ij} are elastic compliances. Both these matrices are symmetric. (Note: in tensor notation we have the equation $\sigma_{ij} = \Sigma\Sigma C_{ijkl} e_{kl}$ with summation over k and l. The inverse is not easily found, which is one reason why the simpler notation is usually preferred. The advantage of the tensor notation comes when calculation of stresses or strains in directions other than the axes of reference are required.)

The matrix **C** has 36 components but in fact these reduce to 21 or less because of thermodynamic considerations and symmetry (see for example [4, 5]). The number of independent elastic constants C_{ij} (or compliances S_{ij}) is 21 when the symmetry is triclinic, 13 when monoclinic and 9 when orthorhombic. Materials with hexagonal or transverse isotropic (fibre) symmetry require five elastic constants. Cubic symmetry requires three and, as seen earlier, isotropic materials need two elastic constants.

Unless there is high symmetry, the relation between the moduli C_{ij} or compliances S_{ij} and more familiar terms such as Young's modulus, E, or the shear modulus, G, is not simple. For isotropic symmetry, however, we need only two elastic constants C_{11} and C_{12}. Then

$$E = C_{11} - \frac{2C_{12}^2}{C_{11} + C_{12}} \quad \text{and} \quad G = \frac{1}{2}(C_{11} - C_{12})$$

Other constants follow by the usual equalities.

For the commonly occurring transverse isotropy, or fibre symmetry, there is a unique direction, the fibre direction, usually chosen to be the 3 axis. Then we may refer to two Young's moduli, namely E_3 and E_1 $(= E_2)$, parallel and transverse to the fibre direction and to two shear moduli, one in a plane containing the 3 axis and one in the plane perpendicular to it. These are, with an obvious notation, referred to as G_{31} $(= G_{23})$ and G_{12} respectively.

For fibre symmetry there are five elastic constants $C_{11}, C_{33}, C_{12}, C_{31}$ and C_{44}. In terms of these constants

$$E_3 = C_{33} - \frac{2C_{31}^2}{C_{11} + C_{12}} = 1/S_{33}$$

$$E_1 = \frac{(C_{11} - C_{12})[C_{33}(C_{11} + C_{12}) - 2C_{31}^2]}{C_{11}C_{33} - C_{31}^2} = 1/S_{11}$$

$$G_{31} = C_{44} = 1/S_{44}$$

$$G_{12} = \frac{1}{2}(C_{11} - C_{12}) = 1/[2(S_{11} - S_{12})]$$

We also have two Poisson ratios ν_{12} and ν_{13} given by

$$\nu_{12} = \frac{-S_{12}}{S_{11}}, \nu_{13} = \frac{-S_{13}}{S_{11}}$$

These are, respectively, the ratios of transverse strain along the 2 or 3 directions divided by longitudinal strain along the 1 direction. The bulk modulus K is given by

$$K = 1/(2S_{11} + 2S_{12} + 4S_{13} + S_{33})$$

6.1.1 Viscoelasticity

We discuss only linear viscoelasticity here and define it as a state in which the time-dependent stress $\sigma(t)$ and the time-dependent strain $e(t)$ are related by a linear differential equation

$$F(D)\sigma = G(D)e$$

where $F(D)$ is the operator

$$A_n \frac{d^n}{dt^n} + A_{n-1} \frac{d^{n-1}}{dt^{n-1}} + \cdots + A_1 \frac{d}{dt} + A_0$$

of nth order and $G(D)$ a corresponding operator

$$E_0 + E_1 \frac{d}{dt} + \cdots + E_m \frac{d^m}{dt^m}$$

of order m.

Simple examples are:

(a) the so-called Maxwell solid where $A_0\sigma + A_1 \dfrac{d\sigma}{dt} = E_1 \dfrac{de}{dt}$

(b) the Voigt–Kelvin solid, where $A_0\sigma = E_0 e + E_1 \dfrac{de}{dt}$

(c) the standard linear solid, where $A_0\sigma + A_1 \dfrac{d\sigma}{dt} = E_0 e + E_1 \dfrac{de}{dt}$

These simple examples are often illustrated by the spring-dashpot models shown in Figure 6.3.

More complex models may be assembled, of course, to represent higher order differential equations. They are usually constructed of parallel or series-parallel assemblies of the simple models. Though giving no information as to the molecular processes responsible for viscoelasticity, they are frequently useful descriptive devices for particular materials.

The solutions of the equations in (a), (b) or (c) may be of two types corresponding to the response of the material to the imposition of a step-function (of stress or of strain) or its response to an oscillatory excitation. Because the equations are linear these solutions are related to each other but that need not concern us here. We shall look at the solutions to equation (c) since they illustrate the main points. Using the model in Figure 6.3, equation (c) becomes:

$$\frac{1}{E_2}\frac{d\sigma}{dt} + \frac{1}{\eta}\left(1 + \frac{E_1}{E_2}\right)\sigma = \frac{de}{dt} + \frac{E_1 e}{\eta}$$

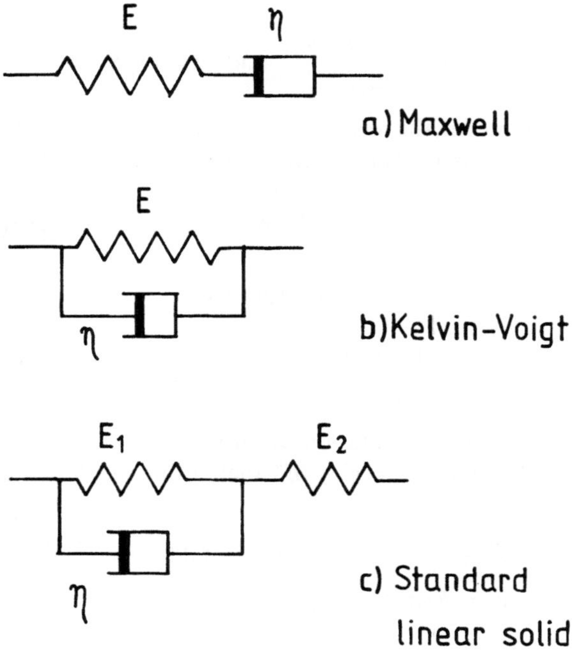

Figure 6.3 Examples of spring-dashpot models. E is a modulus and η the constant of proportionality in linear viscosity.

Under constant strain $e = e_0$, with $\sigma = \sigma_0$ at time $t = 0$, we have, at subsequent times,

$$\sigma(t) = \sigma_\infty + (\sigma_0 - \sigma_\infty)\exp(-t/\tau_1)$$

with

$$\sigma_\infty = \frac{E_2 E_1 e_0}{E_1 + E_2} \quad \text{and} \quad \tau_1 = \eta/(E_1 + E_2).$$

The stress σ therefore decays exponentially to the value σ_∞ with a *relaxation time* τ_1. We can express this equation in a different way by dividing by e_0 and obtaining the time-dependent modulus

$$E(t) = E_R + (E_u - E_R)\exp(-t/\tau_1)$$

with E_R and E_u being the relaxed and unrelaxed moduli respectively:

$$E_R = \frac{E_1 E_2}{E_1 + E_2} \quad \text{and} \quad E_u = E_2$$

In a *creep experiment* $\sigma = \sigma_0$, with $e = e_0$ at time $t = 0$ and

$$e = e_\infty + (e_0 - e_\infty)\exp(-t/\tau_2)$$

Here

$$e_\infty = \frac{(E_2 + E_1)\sigma_0}{E_1 E_2} \quad \text{and} \quad \tau_2 = \eta/E_1.$$

Note that, for this model, the relaxation times for creep and for constant strain are not the same. This is also generally true in experimental work on real materials.

In dynamic strain we apply an oscillatory strain $e = e_1 \sin wt$ and obtain the solution

$$\sigma = A \sin(wt + \delta)$$

where

$$A = e_1 \frac{\tau_1}{\tau_2} E_2 \left(\frac{1 + w^2\tau_2^2}{1 + w^2\tau_1^2} \right)^{\frac{1}{2}} \quad \text{and} \quad \tan\delta = \frac{w(\tau_2 - \tau_1)}{w^2\tau_1\tau_2 + 1}.$$

Again we may express this solution in terms of moduli but these are now complex. First, we define $E^* = E' + iE''$, then, using the definitions of E_u and E_R given earlier, we find

$$E' = E\cos\delta = E_R(1 + w^2\tau_1\tau_2)/(1 + w^2\tau_1^2)$$

$$E'' = E\sin\delta = wE_R(\tau_2 - \tau_1)/(1 + w^2\tau_1^2)$$

Now

$$E_u - E_R = E_u(1 - \tau_1/\tau_2)$$

and therefore

$$E' = E_R + \frac{w^2\tau_1^2(E_u - E_R)}{1 + w^2\tau_1^2}$$

$$E'' = w\tau_1 \frac{(E_u - E_R)}{1 + w^2\tau_1^2}.$$

The storage modulus $|E|$ is given by

$$(E'^2 + E''^2)^{\frac{1}{2}} = \frac{(E_R^2 + E_u^2 w^2\tau_1^2)^{\frac{1}{2}}}{(1 + w^2\tau_1^2)^{\frac{1}{2}}}.$$

The loss factor, $\tan\delta$, is

$$\tan\delta = \frac{w\tau_1(E_u - E_R)}{E_R + w^2\tau_1^2 E_u}$$

6.1.2 Temperature dependence of the relaxation times

The equation for the loss factor, $\tan\delta$, above shows that as frequency is altered the energy lost in (absorbed by) a polymer, varies from a negligible amount at very low frequencies and at very high frequencies to a maximum given by

$$w^2 = \frac{1}{\tau_1^2 E_R E_u}$$

This behaviour is illustrated in Figure 6.4. It is more common to study the temperature dependence of loss factor or modulus rather than the frequency dependence, for experimental reasons, and a typical plot of modulus and loss factor at one frequency as a function of temperature is shown in Figure 6.5 for an epoxy resin system.

The frequency- and the temperature-dependence of both loss and storage moduli are related through the dependence of relaxation times on temperature by the expressions

$$\tau = \tau_0 \exp \frac{\Delta H}{R}\left(\frac{1}{T} - \frac{1}{T_0}\right) \quad \text{(Arrhenius relation)}$$

or

$$\tau = \tau_0 \exp\left(\frac{-C_1(T - T_s)}{C_2 + T - T_s}\right) \quad \text{(Williams–Landel–Ferry relation)}$$

where C_1, C_2 and T_s are constants. For the simpler Arrhenius law the dependence of relaxation time on temperature means that, for example, the compliance (inverse of modulus) at time t in a creep experiment at temperature

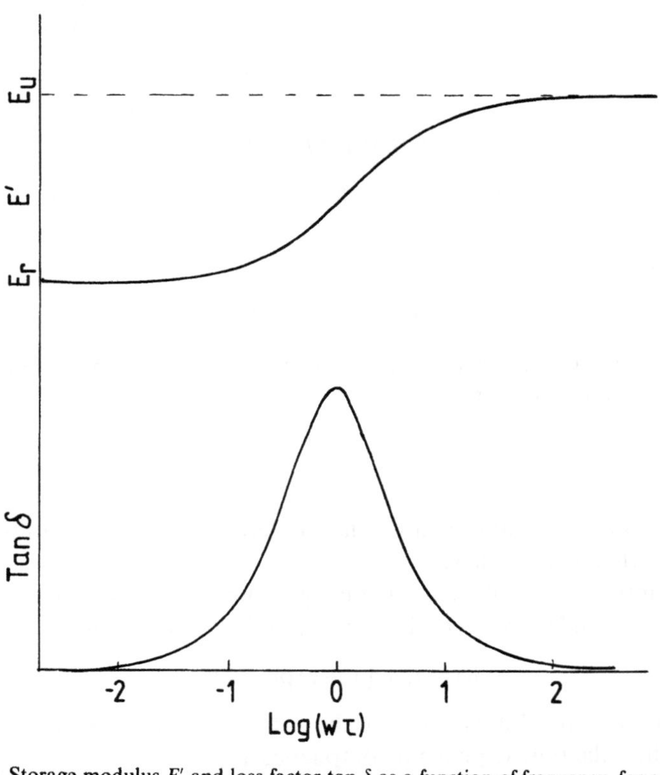

Figure 6.4 Storage modulus E' and loss factor tan δ as a function of frequency, for an idealised single-relaxation-time model.

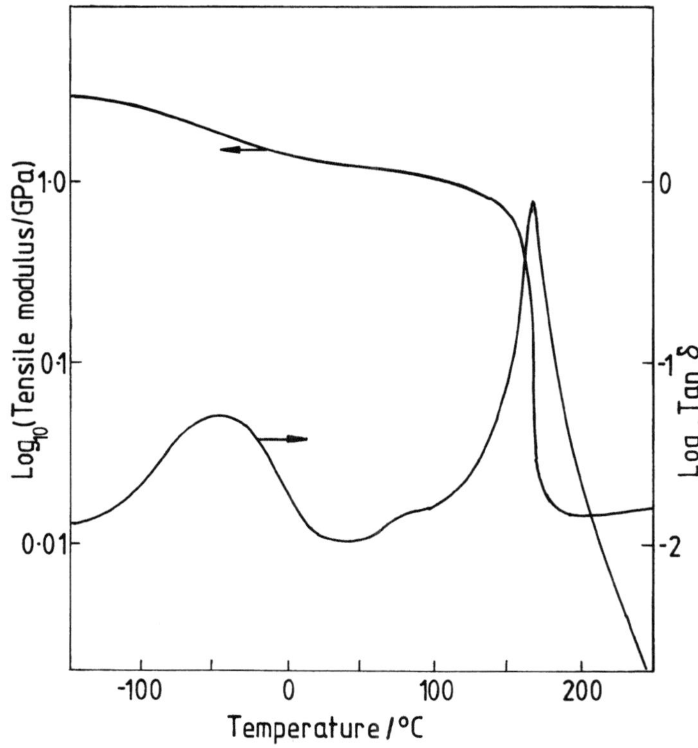

Figure 6.5 Tensile modulus and loss factor as a function of temperature for an epoxy resin system.

T is given by

$$J^T(t) = J^{T_0}(t/a_T)$$

where J^{T_0} is the compliance at a reference temperature T_0, provided that the time is *reduced* by the factor

$$a_T = \frac{t}{t_0}$$

This enables experimental data taken at different temperatures to be superposed to form a master curve.

An example will clarify this time-temperature superposition. Consider a Voigt element under creep conditions, that is, let $\sigma = \sigma_0$, then

$$e(t) = J_0\sigma_0[1 - \exp(-t/\tau)]$$

where σ_0 is the applied stress and J_0 is the instantaneous compliance. Dividing by σ_0 we find the time-dependent compliance as

$$J(t) = J_0[1 - \exp(-t/\tau)]$$

Suppose now that we have determined $J(t)$ over a range of temperatures. We shall have

$$J^{T_0}(t) = J_0^{T_0}\,[1 - \exp(-t/\tau_0)]$$

where τ_0 is the relaxation time relevant at temperature T_0.

At a different temperature T we have

$$\tau = \tau_0 \exp\frac{\Delta H}{R}\left(\frac{1}{T} - \frac{1}{T_0}\right)$$

Write

$$a_T = \frac{\tau}{\tau_0}$$

then

$$J^{T_0}(t) = J_0^{T_0}\{1 - \exp[-t/(\tau/a_T)]\}$$
$$= J_0^{T_0}[1 - \exp(-a_T t/\tau)]$$

which we may write as

$$J_0^{T_0}[1 - \exp(-t'/\tau)]$$

where $t' = a_T t$. However, the compliance at temperature T would normally be written as

$$J^T(t') = J_0^T[1 - \exp(-t'/\tau)]$$

$$= \frac{J_0^T}{J_0^{T_0}} J^{T_0}(t) = \frac{J_0^T}{J_0^{T_0}} J_0^{T_0}(t'/a_T)$$

If the initial compliances J_0^T and $J_0^{T_0}$ are *equal*, then $J^T(t) = J^{T_0}(t/a_T)$, that is, the compliance at a temperature T is the same as that at temperature T_0 but for the *reduced time* t/a_T.

On a log plot

$$J^T(\log t) = J^{T_0}(\log t - \log a_T).$$

(Note: if the compliances J_0^T and $J_0^{T_0}$ are not equal then a scaling of the data by the factor $J_0^T/J_0^{T_0}$ is necessary before the horizontal shifting is done.)

Figures 6.6 to 6.8 show how individual creep curves may superpose under the simple Arrhenius relation to derive a master curve and how, from the value of a_T derived from the shifting, the activation energy ΔH may be derived. For more details see Ferry [2] or McCrum *et al.* [3].

If we insert the Arrhenius expression for the relaxation time in the three-element model analysed earlier we find the storage modulus

$$|E| = \frac{[E_R^2 + E_u^2 w^2 A^2 \exp(2\Delta H/RT)]^{\frac{1}{2}}}{[1 + w^2 A^2 \exp(2\Delta H/RT)]^{\frac{1}{2}}}$$

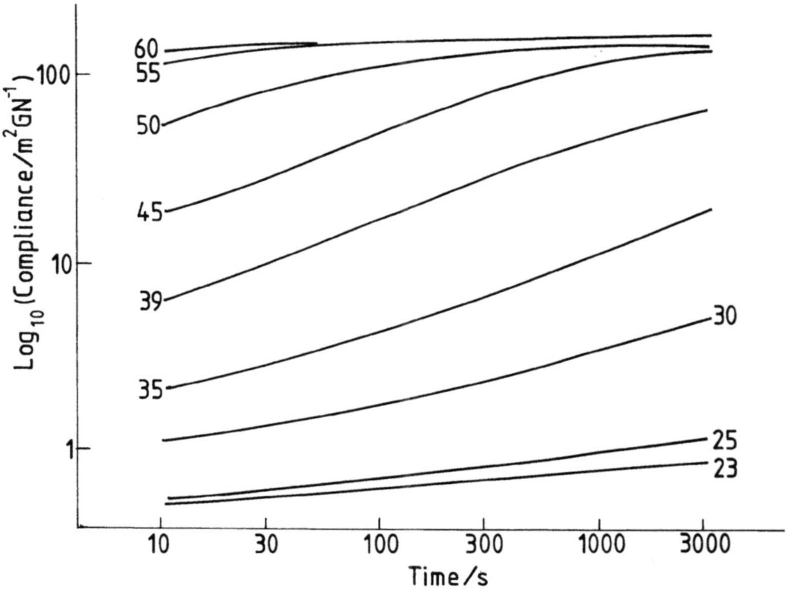

Figure 6.6 Log(compliance) for a flexibilised epoxy resin as a function of time. The numbers on the curves refer to the temperature (°C) of the measurement.

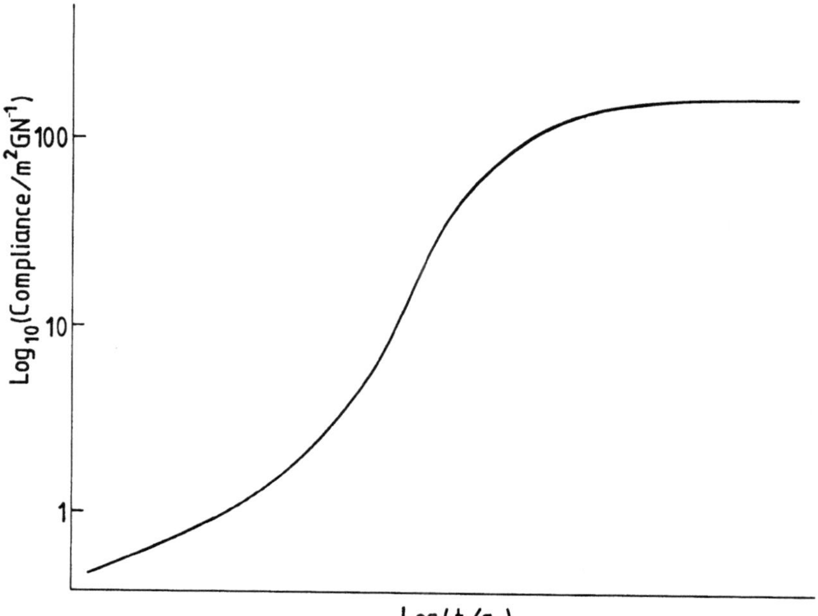

Figure 6.7 Master curve of log(compliance) derived from Figure 6.6 by shifting along the log(time) axis.

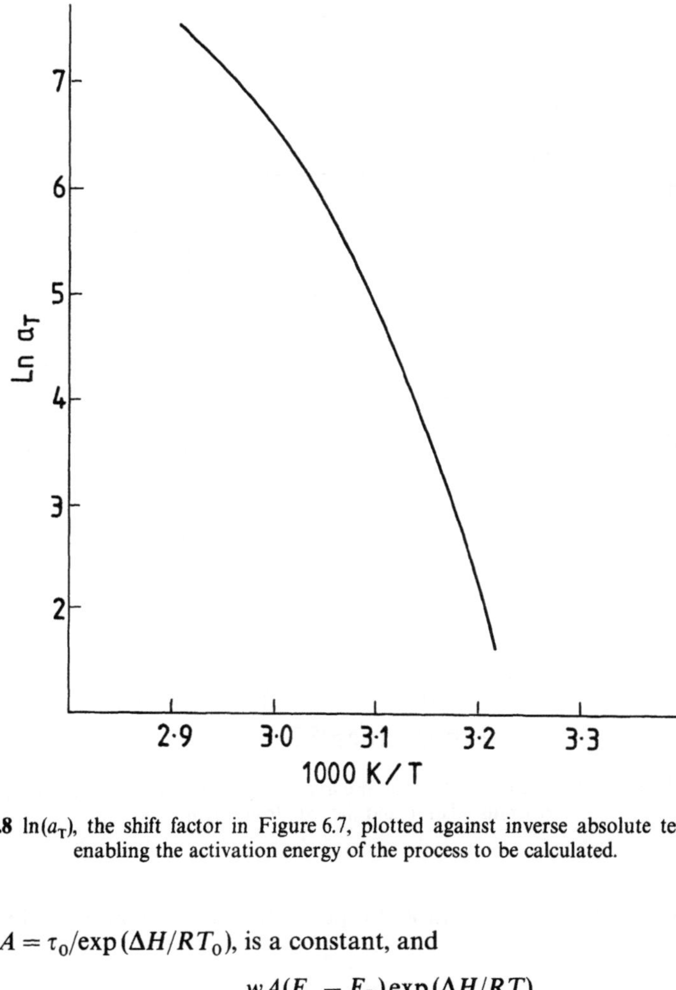

Figure 6.8 $\ln(a_T)$, the shift factor in Figure 6.7, plotted against inverse absolute temperature enabling the activation energy of the process to be calculated.

where $A = \tau_0/\exp(\Delta H/RT_0)$, is a constant, and

$$\tan\delta = \frac{wA(E_u - E_R)\exp(\Delta H/RT)}{E_R + w^2 E_u A^2 \exp(2\Delta H/RT)}$$

The behaviour of $|E|$ and $\tan\delta$ as a function of temperature T is plotted in Figure 6.9. This may be compared with Figure 6.5. The values of the parameters used to produce this plot correspond to the observed values in Figure 6.5. Thus, for the low temperature relaxation $E_u = 3.0$ GPa, $E_R = 1.30$ GPa, $\Delta H = 17.9$ kcal/mole and $T_0 = 228$ K, R, the gas constant, is taken as 1.987×10^{-3} kcal/mole. For the glass transition, T_0 is taken as 433 K, $E_u = 1.30$ GPa and $E_R = 0.0144$ GPa. ΔH for this transition was 160 kcal/mole. The angular frequency $w = 2\pi f$, with $f = 0.67$ Hz. While the fitting of the low temperature relaxation is seen to be good, the glass transition is less well fitted. This is because of the inadequacies of a single-relaxation time model in this region.

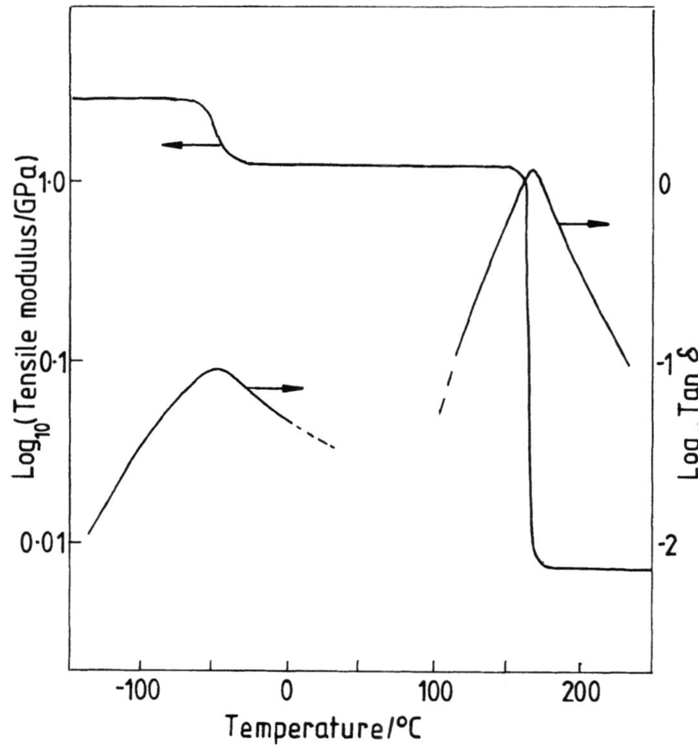

Figure 6.9 Theoretical modulus and loss factor curves using a single-relaxation-time model, the Arrhenius relation and the data from Figure 6.5.

The behaviour of these quantities for polymer blends and alloys is discussed in section 6.7.

6.2 General consideration of the properties of polymer blends and alloys

We now have the techniques necessary to discuss the problems of blends and alloys. From the point of view of their mechanical properties, what concerns us about polymer blends or alloys is the degree to which they are homogeneous. As they depart from the ideal molecular or supermolecular homogeneity, and the dimensions of the regions of inhomogeneity increase, so the blend approaches a composite material in which the phases, now of dimensions of 10 nm or more have clearly different properties. We can then use the terminology of modern composite materials to describe them. In such materials two parameters must be defined. These are: (a) the shapes of the inhomogeneities and (b) their degree of orientation. We may further define a *microcomposite* as a composite in which the dimensions of the phases lie between 10 and 100 nm, a *mesocomposite* as having dimensions between 100 and 1000 nm and a *macrocomposite* as having dimensions greater than 1000 nm, that is, roughly the wavelength of visible light.

As far as the *classical* theory of elasticity is concerned, these dimensions are irrelevant. However, when large strains are involved (rubber elasticity) and when yield and fracture processes are being considered, the dimensions become of vital importance, the physical properties depending strongly on phase dimensions. This fact is made use of in thermoplastic rubbers, typically made from block copolymers in which the dimensions of the constituents place them in the class of mesocomposites. Their behaviour relies on the glassy components acting as crosslinks for the rubbery phase in the manner in which a chemical crosslink would act, but with the advantage that above the glass transition the crosslinks may be removed, to be restored by cooling below the transition after the polymer has been moulded into some new form. By analogy with the behaviour of particles in reinforcing metals, the yield behaviour of copolymers is expected to be similarly influenced and an equation similar to the Hall–Petch relation to hold, defining the yield stress as a function of the dimensions of the harder phase. There does not seem to be very much research along these lines.

(i) Isotropy. A polymer may have isotropic mechanical (and other) properties even if its phases are anisotropic provided that *over a sufficiently small volume* the anisotropy is randomised. Thus if the phases are highly oriented crystallites, a random orientation of microcrystals will result in an isotropic assembly. Of course, if the phases are isotropic and spherically symmetrical this will also ensure overall isotropy (unless the array is ordered into a crystal superlattice). However, if the phases are non-spherical (elliptical, fibrous or lamellar) then even if they are themselves isotropic the differences in elastic (or viscoelastic) properties between phases will result in anisotropy on the global scale unless the geometrical differences are randomised. Schematic diagrams of such assemblages are shown in Figure 6.10.

(ii) Anisotropy. The most elementary form of anisotropy may be termed *molecular anisotropy* which occurs when the normally random arrays of polymer chains become non-random by processing conditions which impart preferred orientation throughout the mass.

The simplest example is the degree of anisotropy (shown by the development of birefringence) caused by large extensions of rubber or rubber-like materials. When glassy polymers such as PMMA or PS are deformed above T_g the anisotropy may be 'frozen-in' when the polymer is cooled below T_g and is visible through birefringence in the solid. What is of more importance in the context of blends and alloys is the preferred orientation brought about in the separate phases either by deformation of a phase from a quasi-spherical into a

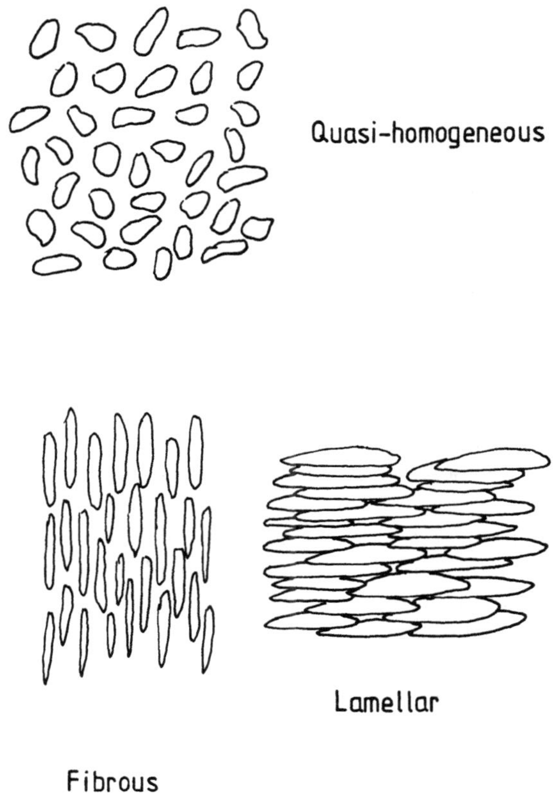

Figure 6.10 Schematic structural elements in a polymer mixture.

non-spherical form or (and in some cases in addition to) orientation of phases along a preferred direction.

If this alteration of the shapes of the phases and their orientation is preserved into the solid state after cooling from the temperature of processing then a permanently anisotropic material results. In the block copolymers of styrene–butadiene–styrene, processing may cause the structure (a mesocomposite) to become rod-like or lamella-like over a large volume, with a consequent effect on the mechanical properties.

Lastly, where the phases are crystalline in, for example, copolymers of ethylene and propylene, processing will cause yield in the crystals and pronounced anisotropy of the material results.

6.3 Simple formulae for the mechanical properties of blends and their limitations

6.3.1 *The 'rule of mixtures'*

The easiest, and sometimes the only way to proceed if little is known about the elastic properties of the phases or of their geometry or orientation is to assume that their effect will be roughly proportional to their volume fraction in the mixture.

Thus 40% of polymer A with modulus E_A, combined with 60% of B, with modulus, E_B will result in a modulus

$$E = 0.4E_A + 0.6E_B$$

Although this has been written for Young's modulus E, it will also apply for the shear modulus G, with as much justification.

This simple rule (which will also apply to the dynamic moduli) must always be treated with great caution. The reasons for its inaccuracy are many and they are discussed fully in the literature on composite materials (see e.g. [6, 7]).

As soon as better information is available as to the nature and properties of the separate phases much better predictions can be made which will be discussed in sections 6.4–6.6. Here, however, mention must be made of another simple way of considering the properties of blends and alloys, the Takayanagi models.

6.3.2 *The Takayanagi models*

These were introduced as long ago as 1963 and represent an improvement on the 'rule of mixtures' law. They have close analogies with the spring-dashpot models for viscoelasticity introduced in section 6.1. Given two materials of elastic moduli E_1 and E_2, they may be arranged in parallel or in series (Figure 6.11a). If their concentrations are C_1 and C_2 respectively, then by considering equal strains in each it is easy to show that the combined modulus is the 'rule of mixtures'

$$E_c = C_1E_1 + C_2E_2.$$

If, however, they are arranged in series (Figure 6.11b), then they carry equal *stress* and we find

$$1/E_c = C_1/E_1 + C_2/E_2.$$

Takayanagi [8] proposed series–parallel models like those in Figure 6.12 where, by considering the combinations separately we find

$$E_{C_a} = \left[\frac{h}{C_1E_1 + C_2E_2} + \frac{1-h}{E_2}\right]^{-1} \text{ and } E_{C_b} = C_1\left(\frac{h}{E_1} + \frac{1-h}{E_2}\right)^{-1} + C_2E_2$$

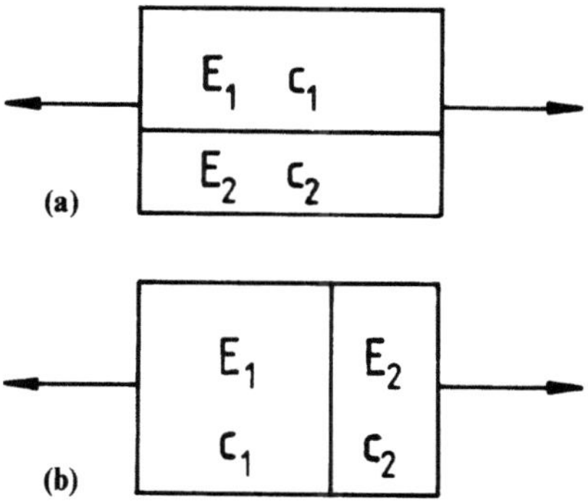

Figure 6.11 The Takayanagi models (a) parallel; (b) series.

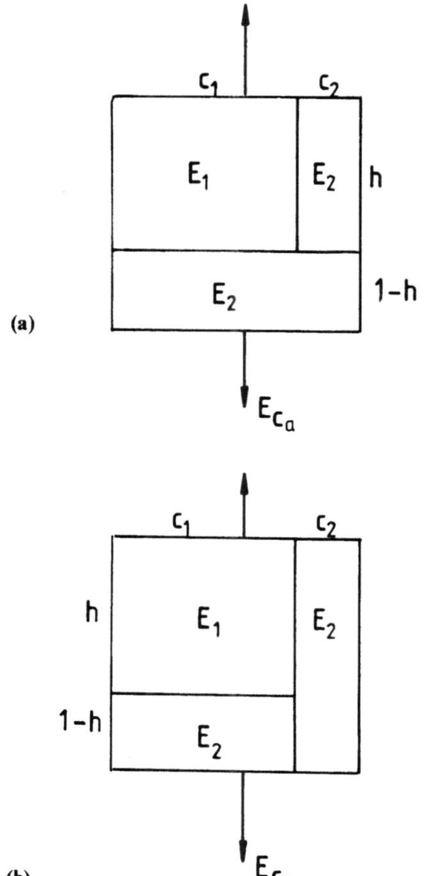

Figure 6.12 The Takayanagi series-parallel models.

In *a* a parallel arrangement of polymers 1 and 2 in the proportions $C_1:C_2$ is in series with a quantity 1-*h* of polymer 2, the proportion of the parallel mixture being *h*. In model *b* a parallel arrangement of pure polymer 1 and a series mixture of 1 and 2 is used. It is tempting to think of these divisions into blocks as real but this should be avoided since important assumptions would have to be made which are not physically realistic. In the first place the assumption made in the Takayanagi scheme is that the stresses and strains are uniform in each block and second, that they are of one form only (usually, as here, tensile).

In a general state of strain in a body there are six components of strain and six of stress and these are related by up to 21 elastic coefficients. Even in the simplest case, where the phases of the polymer 'composite' are isotropic (and of course, in practice, they are not) we have two elastic constants in each phase to consider. Tensile strain will cause lateral contraction according to Poisson's ratio and so, in general, the simple picture given by the Takayanagi models is not exact.

Research in composite materials over recent decades has, however, given clear guidelines on the way to proceed.

6.4 Exact formulae derived for ideal materials

6.4.1 The Voigt and Reuss bounds

In isotropic materials there are two independent elastic constants, the bulk modulus K and the shear modulus G. Young's modulus and Poisson's ratio may be found from these by the formulae

$$E = \frac{9KG}{3K + G} \quad \text{and} \quad \nu = \frac{3K - 2G}{2(3K + G)}$$

Then it can be shown rigorously, using energy arguments, that

$$K_V = C_1K_1 + C_2K_2 \quad \text{and} \quad G_V = C_1G_1 + C_2G_2$$

are upper *bounds* on the bulk and shear moduli respectively of the composite. These are called the Voigt bounds. In a dual argument it can be shown that

$$\frac{1}{K_R} = \frac{C_1}{K_1} + \frac{C_2}{K_2} \quad \text{and} \quad \frac{1}{G_R} = \frac{C_1}{G_1} + \frac{C_2}{G_2}$$

form lower bounds (the Reuss bounds) on the bulk and shear moduli. (In the Voigt bounds it is assumed that a state of uniform *strain* exists in the composite, whereas in the Reuss case there is uniform *stress*.)

Summarising, we have

$$K_r < K_C < K_V \quad \text{and} \quad G_r < G_C < G_R.$$

It is clear that in the Voigt case the stresses cannot be in equilibrium at phase boundaries, whereas in the Reuss case there must be inequalities of strain at the boundaries [6]. In polymers the differences between the moduli of different phases may be so large (particularly where one phase is either glassy or crystalline and the other rubbery) that the overall modulus is nearer to one bound than another. Ward [9] found good agreement with Reuss bounds for polyethylene but in nylon the Voigt average gave the better fit.

6.4.2 *Useful approximations*

It is usual to take Poisson's ratio ν as 0.5 for rubbers and also for most polymers well above their glass transitions. Below T_g, however, ν is usually about 0.35 for most polymers. Again, most polymers have bulk moduli in the region 3–6 GPa (similar to the modulus of a van der Waals solid) regardless of temperature, while their shear moduli may vary from about 1 GPa below T_g to as low as 1 MPa for a lightly crosslinked polymer well above T_g. In the latter case (rubbery modulus) G is linearly proportional to absolute temperature.

6.4.3 *The Hashin–Shtrikman–Hill bounds*

Better bounds (i.e. closer bounds) than the Voigt–Reuss ones may be found by taking, for example, uniform phase stress or phase strain rather than uniform *overall* stress or strain. This leads to the formulae below (the generalisation is due to Walpole [7]).

$$\frac{C_1}{1+(K_1-K_2)C_2/(K_2+K_l^*)} \leqslant \frac{K_C-K_2}{K_1-K_2} \leqslant \frac{C_1}{1+(K_1-K_2)C_2/(K_2+K_g^*)}$$

$$\frac{C_1}{1+(G_1-G_2)C_2/(G_2+G_l^*)} \leqslant \frac{G_C-G_2}{G_1-G_2} \leqslant \frac{C_1}{1+(G_1-G_2)C_2/(G_2+G_g^*)}$$

Here

$$K_l^* = \frac{4}{3}G_l, \qquad K_g^* = \frac{4}{3}G_g \text{ and } G_g^* = \frac{3}{2}\left[\frac{1}{G_g} + \frac{10}{9K_g+8G_g}\right]^{-1}$$

$$G_l^* = \frac{3}{2}\left[\frac{1}{G_l} + \frac{10}{9K_l+8G_l}\right]^{-1}$$

G_g and K_g are the greatest of G_i and K_i in the mixture, while G_l and K_l are the least. (This takes care of the case where, for example $K_1 > K_2$ but $G_1 < G_2$.) If something is known of the geometry of the phases then closer bounds still may be found. When one or more of the phases is anisotropic bounds can also be found for the overall elastic constants. In most blends of polymers this does not apply but reference may be made to Walpole [10] for the relevant theory.

6.5 Special relations for particular geometries

6.5.1 Arrays of fibres

Processing may cause the phases to become aligned in a preferred direction. The dependence of elastic properties on the degree of orientation will be discussed in the next section but here it will be assumed (i) that the orientation is near-perfect and (ii) that the length/diameter ratio (aspect ratio) of the fibres is very large. The polymer has then become a micro- or meso-composite and will have noticeably anisotropic properties overall. For example, in a styrene–butadiene–styrene (SBS) block copolymer, the polystyrene forms cylinders of 15 nm diameter in hexagonal packing with a unit cell dimension of 30 nm. In such materials Young's modulus at any angle θ to the axis of symmetry can be found from the formula

$$E_\theta^{-1} = S_{11}\sin^4\theta + (2S_{13} + S_{44})\sin^2\theta\cos^2\theta + S_{33}\cos^4\theta$$

where the elastic compliances S_{ij} may be derived from classical fibre reinforcement theory if the elastic properties of the two phases are known.

Thus [11]:

$$S_{11} = S_{22} = \frac{(1-\nu_2)^2}{E_2}\left[\frac{2+(a_1-1)G_2/G_1}{2-c+ca_2+(1-c)(a_1-1)G_2/G_1} - \frac{2c(1-G_2/G_1)}{c+a_2+(1-c)G_2/G_1}\right]$$

$$S_{44} = S_{55} = \frac{1-c+(1+c)G_2/G_1}{1+c+(1-c)G_2/G_1}\cdot\frac{1}{G_2}$$

$$S_{66} = \frac{(1-c)a_2+(1+ca_2)G_2/G_1}{c+a_2+(1-c)G_2/G_1}\cdot\frac{1}{G_2}$$

$$S_{23} = S_{32} \sim S_{13} = S_{31} = -S_{22}\left[\nu_2 - \frac{4c(1-\nu_2)(\nu_2-\nu_1)}{2-c+ca_2+(1-c)(a_1-1)G_2/G_1}\right]$$

$$S_{33}^{-1} = cE_1 + (1-c)E_2 + \frac{8c(1-c)G_2(\nu_1-\nu_2)^2}{2-c+ca_2+(1-c)(a_1-1)G_2/G_1}$$

$$S_{12} = S_{21} = S_{33}\left[\nu_2 - \frac{c(1+a_2)(\nu_2-\nu_1)}{2-c+ca_2+(1-c)(a_1-1)G_2/G_1}\right]^2 - \left(\frac{1+\nu_2}{E_2}\right)\left\{\nu_2 + (1-\nu_2)c\left[\frac{a_2-1-(a_1-1)G_2/G_1}{2-c+ca_2+(1-c)(a_1-1)G_2/G_1} - \frac{2(1-G_2/G_1)}{c+a_2+(1-c)G_2/G_1}\right]\right\}$$

In these expressions E_i, G_i, ν_i are respectively Young's and shear moduli and Poisson's ratio for phase i, $a_i = 3 - 4\nu_i$ and

$$S_{12} = -\nu_{12}S_{11} = S_{21} = -\nu_{21}S_{22}$$

$$S_{23} = -\nu_{23}S_{22} = S_{32} = -\nu_{32}S_{33}$$
$$S_{31} = -\nu_{31}S_{33} = S_{13} = -\nu_{13}S_{11}.$$

6.5.2 *Stacks of lamellae*

Processing of SBS block copolymers can also lead to formation of a micro- or meso-composite with lamellar structure. This will, like an aligned fibre composite also have transverse isotropy with five elastic constants. The C_{ij} for a lamella stack were first derived by Bruggeman [12] and may be inverted to give the S_{ij}. The S_{ij} are derived explicitly by Allan *et al.* [13] as follows.

$$E_{90} = S_{11}^{-1} = E_{c_1} = c_1E_1 + c_2E_2 - \frac{c_1c_2E_1E_2(\nu_1-\nu_2)^2}{c_1E_1(\nu_2^2-1)+c_2E_2(\nu_1^2-1)}$$

$$S_{22} = S_{11}$$

$$E_0^{-1} = S_{33} = \frac{c_1}{E_1} + \frac{c_2}{E_2} - \frac{2c_1c_2(\nu_2E_1-\nu_1E_2)^2}{E_1E_2[c_1E_1(1-\nu_2)+c_2E_2(1-\nu_1)]}$$

$$S_{44} = S_{55} = \frac{c_1}{G_1} + \frac{c_2}{G_2}$$

$$S_{66} = 2(S_{11} - S_{12}) = (c_1G_1 + c_2G_2)^{-1}$$

$$S_{12} = S_{11} - S_{66}/2$$

$$\frac{1}{K_C} = \frac{c_1}{E_1} + \frac{c_2}{E_2} - 2\left(\frac{\nu_2}{E_2} + \frac{\nu_1}{E_1}\right) + \frac{2\{c_2[(\nu_2/E_2)-(\nu_1/E_1)]+(\nu_2-1)/E_2\}\{c_1[(\nu_1/E_1)-(\nu_2/E_2)]+(\nu_1-1)/E_1\}}{c_2(1-\nu_1)/E_1 + c_1(1-\nu_2)/E_2}$$

$$S_{13} = \frac{1}{4}\left(\frac{1}{K_C} - S_{33} - 2S_{11} - 2S_{12}\right)$$

$$S_{23} = S_{13}$$

All other constants are zero

6.5.3 *Array of spheres*

In a cubic array of spheres, which may be approximated by certain copolymers, there are three elastic constants C_{11}, C_{12} and C_{44}. These may be expressed in terms of the elastic properties of the phases, the type of symmetry (simple cubic, body-centred or face-centred) and the concentration of spheres. We have

$$C_{11} = \lambda_2 + G_2(2 + 2\alpha + \gamma)$$
$$C_{12} = \lambda_2 + G_2\gamma$$
$$C_{44} = G_2(1 + \beta)$$

Here

$$\lambda_i = K_i - 2/3G_i$$

and α, β and γ depend on the elastic constants λ_1 and μ_1 of the spheres, their concentration and the symmetry. Sangani and Lu [14] give curves showing the dependence of α, β and γ on concentration for various combinations of the shear moduli of the phases. Arridge [15] gives expressions for the bulk modulus and thermal expansion coefficient of composites made of spherical particles in body-centred cubic or face-centred cubic arrays.

6.6 Orientation

If the effect of processing has been to induce partial orientation of the polymer then the cases discussed in section 6.5 are not relevant and a description of the degree of orientation which may vary from point to point of the material

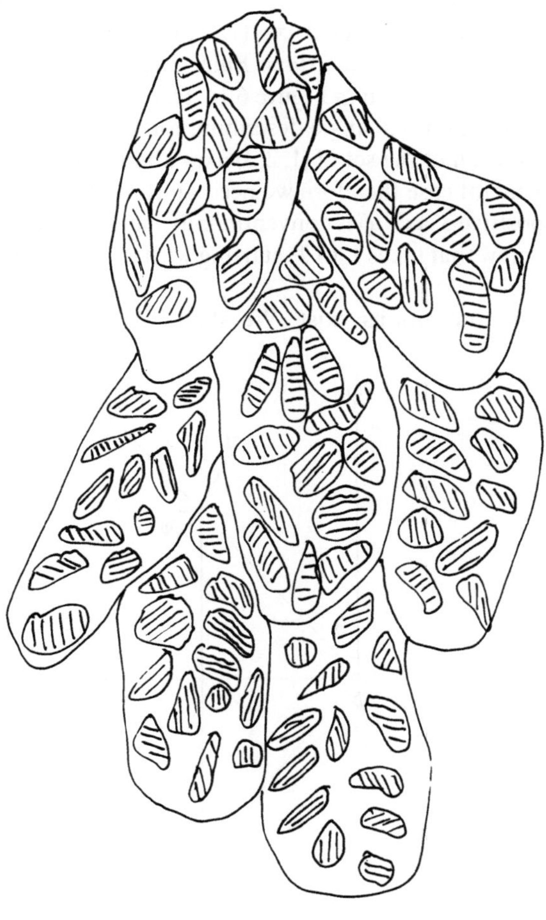

Figure 6.13 Schematic hierarchical scheme of elements in a mixture.

becomes necessary. This may become a matter of considerable complication and require extensive analysis for a proper understanding. However, if we can find expressions for the overall elastic constants of a *representative volume fraction* (RVF) in terms of those of its elements, their concentration and the distribution of their orientations then assemblies of such unit cells, if *randomly* oriented, will have isotropic elastic properties which can be calculated, whereas if the unit cells themselves are preferentially oriented a re-application of the theory will yield a new set of elastic constants for the assembly. 'Hierarchies' of orientation can then be dealt with (see Figure 6.13).

6.6.1 *Local and global axes for a representative volume fraction*

For the most general treatment of orientation we need to consider the description of the elements of the RVF in that volume. They will, in general, be anisotropic with elastic constants C_{ij} (or compliances S_{ij}) referred to a set of local axes. The relation of these local axes (1, 2, 3) to the global axes (1′, 2′, 3′) of the RVF may be described by the three Euler angles θ, ϕ, ψ in the spherical polar coordinate system shown in Figure 6.14. Here θ is the angle between the 3 and 3′ axes, ϕ the azimuthal angle describing rotation about the 3′ axis and ψ the angle of rotation of the set of (1, 2, 3) axes about the local 3 axis.

In the most general case therefore we need all three angles. For θ the zero position is the positive 3′ axis. ϕ is conventionally measured in the direction of a right handed screw that takes 1′ to 2′, starting with $\phi = 0$ along the positive 1′ direction.

For ψ various conventions exist. We shall take another right-handed

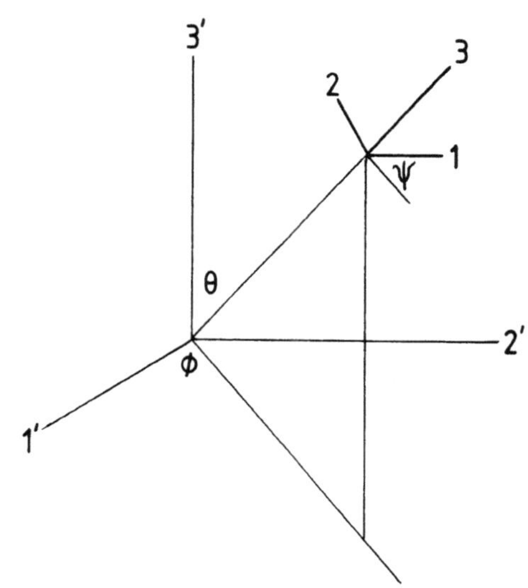

Figure 6.14 Local and global systems of axes.

convention as shown in Figure 6.10 taking $\psi = 0$ in the direction, lying in the 3′3 plane, of increasing θ.

The direction of each of the axes 1, 2 and 3, relative to the global axes 1′ 2′ 3′ is given by three direction cosines $l_{i'j}$. These are given in the appendix to this chapter.

In the global axes the elastic constants $C_{i'j'}$ (or $S_{i'j'}$) are related to those in the local axes by a set of transformation equations. These are best expressed succinctly by the *tensor* equations

$$C_{i'j'k'l'} = l_{i'i} l_{j'j} l_{k'k} l_{l'l} C_{ijkl}$$

(and a similar equation for the S_{ijkl}). In these equations there is summation over the i, j, k and l. The relations required are also given in the appendix.

Knowing the global values of the elastic constants we can form Voigt or Reuss sums to give the overall mean elastic constants, provided we know $\Phi(\theta, \phi, \psi)$, i.e. the *orientation distribution function.* This is defined as follows: the number of elements with axes (1, 2, 3) lying in the solid angle

$$d\Omega = (\theta, \theta + d\theta; \phi, \phi + d\phi; \psi, \psi + d\psi)$$

is given by

$$N\Phi(\theta, \phi, \psi) \sin\theta \, d\theta \, d\phi \, d\psi.$$

The term $\sin\theta$ enters because the element of surface area in the spherical polar coordinates is $\sin\theta \, d\theta \, d\phi$. The total number of elements is then integrated over θ, ϕ, and ψ giving

$$N \int_0^{2\pi} d\psi \int_0^{2\pi} d\phi \int_0^{\pi} \Phi(\theta, \phi, \psi) \sin\theta \, d\theta$$

Normalising, so that this total number is unity we find that the probability of an element with orientations in $d\Omega$ is

$$\frac{1}{8\pi^2} \Phi(\theta, \phi, \psi) \sin\theta \, d\theta \, d\phi \, d\psi.$$

Then the mean Voigt value of $C_{i'j'k'l'}$ is

$$\langle C_{i'j'k'l'} \rangle = \frac{1}{8\pi^2} \iiint l_{i'i} l_{j'j} l_{k'k} l_{l'l} \Phi(\theta, \phi, \psi) C_{ijkl} \sin\theta \, d\theta \, d\phi \, d\psi$$

and similarly for the Reuss average $\langle S_{i'j'k'l'} \rangle$.

In general these averages are difficult to work out unless there is a fair amount of symmetry present. If the orientation is *random* then $\Phi(\theta, \phi, \psi) = 1$ and we can find, after a little effort,

$$E_V = \frac{(A - B + 3C)(A + 2B)}{2A + 3B + C}, \quad E_R = \frac{5}{3A' + 2B' + C'}$$

$$G_V = \frac{A - B + 3C}{5}, \quad G_R = \frac{5}{4A' - 4B' + 3C'}$$

where

$$3A = C_{11} + C_{22} + C_{33}$$
$$3B = C_{23} + C_{31} + C_{12}$$
$$3C = C_{44} + C_{55} + C_{66}$$

and

$$3A' = S_{11} + S_{22} + S_{33}$$
$$3B' = S_{23} + S_{31} + S_{12}$$
$$3C' = S_{44} + S_{55} + S_{66}$$

The next most useful symmetry to consider is that of transverse or fibre symmetry where there are five elastic constants

$$C_{11} = C_{22}, C_{33}, C_{44} = C_{55}, C_{31} = C_{23} \text{ and } C_{12}$$

C_{66} is found from the equation

$$C_{66} = (C_{11} - C_{12})/2.$$

Transverse isotropy of the elements means that integration over ψ introduces a factor 2π. If the global symmetry is also that of transverse isotropy then integration may usually be done over ϕ introducing another factor 2π.

For local and global transverse isotropy only θ is a variable and we have $\frac{\Phi(\theta)}{2}\sin\theta\, d\theta$ as the probability that the 3 direction of the element lies in the *solid cone* of angle $(\theta, \theta + d\theta)$ (Figure 6.15). In such cases we may find simple expressions for the $\langle C_{i'j'} \rangle$. For example:

$$\langle C_{1'1'} \rangle = \frac{1}{2}\int_0^\pi \sin\theta\,\Phi(\theta)\,d\theta \times \Big\{C_{11} + \sin^2\theta(C_{13} + 2C_{44} - C_{11})$$
$$+ \frac{3}{8}\sin^4\theta(C_{33} + C_{11} - 2C_{13} - 4C_{44})\Big\}$$

When there is global fibre symmetry it is convenient to describe the distribution function $\Phi(\theta)$ in terms of an expansion of Legendre polynomials

$$\Phi(\theta) = \sum a_n P_n(\cos\theta)$$

where

$$a_n = \frac{2n+1}{2}\int_0^\pi \Phi(\theta)\,P_n(\cos\theta)\sin\theta\,d\theta$$
$$= \frac{2n+1}{2}\langle P_n(\cos\theta)\rangle$$

For mechanical properties only two of the $P_n(\cos\theta)$ are needed.

$$P_2(\cos\theta) = \frac{1}{2}(3\cos^2\eta - 1)$$

and

$$P_4(\cos\theta) = \frac{1}{8}[35\cos^4\theta - 30\cos^2\theta + 3].$$

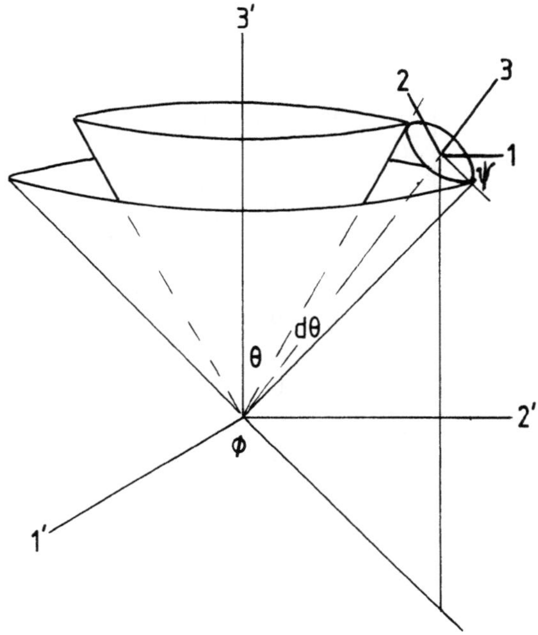

Figure 6.15 How the local axes in Figure 6.14 lie on a cone when transverse anisotropy (fibre symmetry) is present.

The quantities $\langle P_2 \rangle$ and $\langle P_4 \rangle$ may be found experimentally by techniques such as optical birefringence, nuclear magnetic resonance spectroscopy and infrared dichroism (see, for example [16]). We may then write the $\langle C_{i'j'} \rangle$ compactly in matrix form as $\mathbf{a}^T\mathbf{MC}$.

Thus

$$\langle C_{1'1'} \rangle = (a_0\, a_2\, a_4) \begin{bmatrix} \frac{8}{15} & \frac{1}{5} & 0 & \frac{4}{15} & \frac{8}{15} \\ \frac{8}{105} & \frac{-2}{35} & 0 & \frac{-2}{105} & \frac{-4}{105} \\ \frac{1}{105} & \frac{1}{105} & 0 & \frac{-2}{105} & \frac{-4}{105} \end{bmatrix} \begin{bmatrix} C_{11} \\ C_{33} \\ C_{12} \\ C_{31} \\ C_{44} \end{bmatrix}$$

For $\langle C_{1'2'} \rangle$ the matrix is

$$\begin{bmatrix} \frac{1}{15} & \frac{1}{15} & \frac{1}{3} & \frac{8}{15} & \frac{-4}{15} \\ \frac{-2}{105} & \frac{-2}{105} & \frac{2}{15} & \frac{-2}{21} & \frac{8}{105} \\ \frac{1}{315} & \frac{1}{315} & 0 & \frac{-2}{315} & \frac{-4}{315} \end{bmatrix}$$

For $\langle C_{3'1'}\rangle$ it is

$$\begin{bmatrix} \frac{1}{15} & \frac{1}{15} & \frac{1}{3} & \frac{8}{15} & \frac{-4}{15} \\ \frac{1}{105} & \frac{1}{105} & \frac{-1}{15} & \frac{1}{21} & \frac{-4}{105} \\ \frac{-4}{315} & \frac{-4}{315} & 0 & \frac{8}{315} & \frac{16}{315} \end{bmatrix}$$

For $\langle C_{3'3'}\rangle$

$$\begin{bmatrix} \frac{8}{15} & \frac{1}{5} & 0 & \frac{4}{15} & \frac{8}{15} \\ \frac{-16}{105} & \frac{4}{35} & 0 & \frac{4}{105} & \frac{8}{105} \\ \frac{8}{315} & \frac{8}{315} & 0 & \frac{-16}{315} & \frac{-32}{315} \end{bmatrix}$$

and for $\langle C_{4'4'}\rangle$

$$\begin{bmatrix} \frac{7}{30} & \frac{1}{15} & \frac{-1}{6} & \frac{-2}{15} & \frac{2}{5} \\ \frac{-1}{42} & \frac{1}{105} & \frac{1}{30} & \frac{-2}{105} & \frac{1}{35} \\ \frac{-4}{315} & \frac{-4}{315} & 0 & \frac{8}{315} & \frac{16}{315} \end{bmatrix}$$

If the distribution of unit cells is *random* then $a_0 = 1, a_2 = a_4 = 0$ and the $\langle C_{i'j'}\rangle$ are given by the first lines only of the matrices after multiplication. Thus

$$\langle C_{1'1'}\rangle = \frac{8}{15}C_{11} + \frac{1}{5}C_{33} + \frac{4}{15}C_{31} + \frac{8}{15}C_{44}$$

For an isotropic assembly only two elastic constants are required $\langle C_{1'1'}\rangle$ and $\langle C_{1'2'}\rangle$. For the above random assembly

$$\langle C_{1'2'}\rangle = \frac{1}{15}C_{11} + \frac{1}{15}C_{33} + \frac{1}{3}C_{12} + \frac{8}{15}C_{31} - \frac{4}{15}C_{44}$$

Knowing these, the more familiar Young's and shear moduli may be found, since

$$E = C_{11} - \frac{2C_{12}^2}{C_{11} + C_{12}}, \quad G = \frac{1}{2}(C_{11} - C_{12}), \quad \nu = \frac{C_{12}}{C_{11} + C_{12}}$$

It is from these formulae that the Voigt averages for the random assembly, given earlier, have been derived.

For fully aligned elements we have $a_0 = 1, a_2 = 5$ and $a_4 = 9$. To see this consider, for example

$$a_2 = \frac{5}{2}\int_0^{\pi} \Phi(\theta)\, P_2(\cos\theta) \sin\theta\, d\theta$$

Now the probability that θ lies within the *cone* $(\theta, \theta + d\theta)$ is given by

$$\frac{1}{2}\Phi(\theta)\sin\theta\, d\theta$$

so that

$$a_2 = 5\langle P_2(\cos\theta)\rangle$$

the average being taken over the *entire sphere.* Then $\langle P_2(\cos\theta)\rangle = 1$ and so $a_2 = 5$. A similar argument gives $a_4 = 9$. For an equatorial sheet (all elements lying in a plane perpendicular to the 3 axis) we have

$$a_0 = 1, \quad a_2 = -\frac{5}{2}, \quad a_4 = \frac{27}{8}$$

For other orientations, when a_2 and a_4 have been found experimentally, the mean elastic constants $\langle C_{i'j'}\rangle$ may be found from the matrix equations given above. (Note: the derivation of $\langle P_2\rangle$ from birefringence measurements is simple if the birefringence for perfect orientation is known. Calling this Δn_{max} and the measured birefringence for elements with orientation distribution $\Phi(\theta)$, Δn then $\Delta n = \langle P_2(\cos\theta)\rangle \Delta n_{max}$ and $a_2 = 5\langle P_2\rangle$.)

If the compliances $\langle S_{i'j'}\rangle$ are used to find a Reuss average the matrices dual to those given above for the $\langle C_{i'j'}\rangle$ are:

for $\langle S_{1'1'}\rangle$,

$$\begin{bmatrix} \frac{8}{15} & \frac{1}{5} & 0 & \frac{4}{15} & \frac{8}{15} \\ \frac{8}{105} & \frac{-2}{35} & 0 & \frac{-2}{105} & \frac{-4}{105} \\ \frac{1}{105} & \frac{1}{105} & 0 & \frac{-2}{105} & \frac{-4}{105} \end{bmatrix}$$

for $\langle S_{1'2'}\rangle$,

$$\begin{bmatrix} \frac{1}{15} & \frac{1}{15} & \frac{1}{3} & \frac{8}{15} & \frac{-1}{15} \\ \frac{-2}{105} & \frac{-2}{105} & \frac{2}{15} & \frac{-2}{21} & \frac{2}{105} \\ \frac{1}{315} & \frac{1}{315} & 0 & \frac{-2}{315} & \frac{-1}{315} \end{bmatrix}$$

for $\langle S_{3'1'}\rangle$,

$$\begin{bmatrix} \frac{1}{15} & \frac{1}{15} & \frac{1}{3} & \frac{8}{15} & \frac{-1}{15} \\ \frac{1}{105} & \frac{1}{105} & \frac{-1}{15} & \frac{1}{21} & \frac{-1}{105} \\ \frac{-4}{315} & \frac{-4}{315} & 0 & \frac{8}{315} & \frac{4}{315} \end{bmatrix}$$

for $\langle S_{3'3'} \rangle$

$$\begin{bmatrix} \frac{8}{15} & \frac{1}{5} & 0 & \frac{4}{15} & \frac{2}{15} \\ \frac{-16}{105} & \frac{4}{35} & 0 & \frac{4}{105} & \frac{2}{105} \\ \frac{8}{315} & \frac{8}{315} & 0 & \frac{-16}{315} & \frac{-8}{315} \end{bmatrix}$$

and for $\langle S_{4'4'} \rangle$

$$\begin{bmatrix} \frac{14}{15} & \frac{4}{15} & \frac{-2}{3} & \frac{-8}{15} & \frac{2}{5} \\ \frac{-2}{21} & \frac{4}{105} & \frac{2}{15} & \frac{-8}{105} & \frac{1}{35} \\ \frac{-16}{315} & \frac{-16}{315} & 0 & \frac{32}{315} & \frac{16}{315} \end{bmatrix}$$

Using the methods of this section and the S_{ij} found for the fibre or lamella 'unit crystal' forms of SBS block copolymers and referred to in section 6.5 the elastic constants of, for example, injection moulded specimens of these materials may be predicted, if their global pattern of orientation is known.

Example. Allen *et al.* [13] found the following values for a transversely isotropic lamellar SBS copolymer, with 46% polystyrene and 54% polybutadiene.

$$\begin{aligned} S_{11} &= 1.083 \times 10^{-9}\ \mathrm{N^{-1}m^2} \\ S_{33} &= 29.4 \times 10^{-9}\ \mathrm{N^{-1}m^2} \\ S_{44} &= 264 \times 10^{-9}\ \mathrm{N^{-1}m^2} \\ S_{12} &= -0.358 \times 10^{-9}\ \mathrm{N^{-1}m^2} \\ S_{13} &= -0.471 \times 10^{-9}\ \mathrm{N^{-1}m^2} \end{aligned}$$

Inserting these values in the expression for the Reuss average of a randomly ordered assembly of such 'unit cells' we find $\langle S_{1'1'} \rangle = 147.1 \times 10^{-9}\ \mathrm{N^{-1}m^2}$ and $\langle S_{1'2'} \rangle = -15.94 \times 10^{-9}\ \mathrm{N^{-1}m^2}$, from which $E_R = 6.76 \times 10^6\ \mathrm{N\,m^{-2}}$ and $G_R = 3.06 \times 10^6\ \mathrm{N\,m^{-2}}$. We may also invert the S_{ij} to find:

$$\begin{aligned} C_{11} &= 1051.3 \times 10^6\ \mathrm{N\,m^{-2}} \\ C_{33} &= 34.74 \times 10^6\ \mathrm{N\,m^{-2}} \\ C_{44} &= 3.79 \times 10^6\ \mathrm{N\,m^{-2}} \\ C_{12} &= 357.3 \times 10^6\ \mathrm{N\,m^{-2}} \\ C_{13} &= 22.6 \times 10^6\ \mathrm{N\,m^{-2}} \end{aligned}$$

Then $\langle C_{1'1'} \rangle = 575.5 \times 10^6\ \mathrm{N\,m^{-2}}$ and $\langle C_{1'2'} \rangle = 203.3 \times 10^6\ \mathrm{N\,m^{-2}}$ are the Voigt averages and we find $E_V = 469.4 \times 10^6\ \mathrm{N\,m^{-2}}$ and $G_V = 186.1 \times 10^6\ \mathrm{N\,m^{-2}}$.

Such a large difference between the Voigt and Reuss bounds is typical when there are large differences between the elastic constants of the components of the composite, or when there is considerable anisotropy. These results should be compared with the remarks made earlier about the Voigt and Reuss bounds for polyethylene and for nylon.

For comparison with the above results we may also calculate the Voigt and Reuss limits on E and G directly for a composite consisting of 46% polystyrene and 54%. We find

$$E_V = 921.5 \times 10^6\,\mathrm{N\,m^{-2}}, \quad G_V = 346.1 \times 10^6\,\mathrm{N\,m^{-2}}$$
$$E_R = 10.91 \times 10^6\,\mathrm{N\,m^{-2}}, \quad G_R = 3.79 \times 10^6\,\mathrm{N\,m^{-2}}$$

It is interesting to note that the Reuss limits for the composite of arbitrary geometry are close to those for the randomised lamellae, while the Voigt limits for the randomised lamellae are about one half of the values for the generalised geometry.

6.7 Temperature dependence of mechanical properties

Much of what has been described in sections 6.3–6.6 applies when time-dependent moduli are inserted in the equations, provided that the strains are not too large so that linear viscoelasticity is a resonable approximation. It is usual to substitute complex moduli $E^* = E' + iE''$ for the real moduli and then separate out real and imaginary parts. Let

$$E_1^* = E_{1R} + \frac{w^2\tau_1^2(E_{1U} - E_{1R})}{1 + w^2\tau_1^2} + i\left[\frac{w\tau_1(E_{1U} - E_{1R})}{1 + w^2\tau_1^2}\right]$$

$$E_2^* = E_{2R} + \frac{w^2\tau_2^2(E_{2U} - E_{2R})}{1 + w^2\tau_2^2} + i\left[\frac{w\tau_2(E_{2U} - E_{2R})}{1 + w^2\tau_2^2}\right]$$

Then in a simple rule-of-mixtures model we would have $E_c^* = C_1E_1^* + C_2E_2^*$ and so

$$E_c' = C_1E_{1R} + C_2E_{2R} + w^2\left[\frac{C_1\tau_1^2(E_{1U} - E_{1R})}{1 + w^2\tau_1^2} + \frac{C_2\tau_2^2(E_{2U} - E_{2R})}{1 + w^2\tau_2^2}\right]$$

and

$$E_c'' = w\left[\frac{C_1\tau_1(E_{1U} - E_{1R})}{1 + w^2\tau_1^2} + \frac{C_2\tau_2(E_{2U} - E_{2R})}{1 + w^2\tau_2^2}\right]$$

from which $\tan\delta_c = E_c''/E_c'$ can be found.

To use the Voigt and Reuss bounds or the Hashin–Shtrikman–Hill bounds we would need expressions for the two *complex* moduli K and G of the components of the mixture. These may be available from experiment or may be approximated by three- or multielement spring-dashpot models as follows. The 'elements' in Figure 6.3 for the three-element model are stiffnesses and viscosities. These are just as appropriate (though, of course, will not be identical) whether considered as tensile, shear or bulk stiffnesses or as tensile, shear or bulk viscosities. We may therefore postulate a simple three-element bulk modulus

$$K^* = K_R + \frac{w^2\tau_K^2(K_U - K_R)}{1 + w^2\tau_K^2} + i\frac{w\tau_K(K_U - K_R)}{1 + w^2\tau_K^2}$$

K_U and K_R are the unrelaxed and relaxed bulk moduli and τ_K is a bulk relaxation time. Similarly we may define a complex shear modulus

$$G^* = G_R + \frac{w^2\tau_G^2(G_U - G_R)}{1 + w^2\tau_G^2} + i\frac{w\tau_G(G_U - G_R)}{1 + w^2\tau_G^2}$$

These quantities may be assigned to each of the components in the composite and complex Voigt and Reuss bounds or Hashin–Shtrikman–Hill bounds determined. Such exercises may be a useful first approximation to the complete explanation of the behaviour of polymer composites.

Lastly if, as in section 6.1, we insert the Arrhenius expression or the Williams–Landel–Ferry one for the relaxation times in the above we shall predict the behaviour of the composite as a function of temperature. Each component having its own relaxation time (and not necessarily the same one in bulk and in shear), with its characteristic activation energy, we will expect to obtain the behaviour shown in Figure 6.16 where phase miscibility is not perfect and the separate glass transition temperatures are clearly distinguished. Such behaviour has been widely observed and reported in the literature (see, for example [8, 17, 18]). The presence or absence of separate glass transition peaks is used as a measure of miscibility. We can see from the foregoing analysis that it is also a measure of the mechanical hetero- or homogeneity of the mixture.

In some polymer blends the mixing is at molecular level and here the phase separation is not sufficient to allow the separate phases to behave as they would in isolation. One phase affects the other in such cases and the position of T_g is described by the empirical Gordon–Taylor equation, T_g varying linearly between the values for the two phases according to the concentration of each in the mixture (Figure 6.17). It is commonly written as

$$C_1(T_g - T_{g1}) + K(1 - C_1)(T_g - T_{g2}) = 0$$

where C_1 is the concentration of component 1 and K is a constant. K is given the meaning $W_1\alpha_1/W_2\alpha_2$ by diMarzio and Gibbs [19] where W_1 and W_2 are the molecular weights of components 1 and 2, while α_1 and α_2 are the respective numbers of flexible units in the two components.

In a modification of the Gordon–Taylor equation given by Schneider [20] the glass transition of the mixture, T_g, is given by the equation

$$(T_g - T_{g1})/(T_{g2} - T_{g1}) = (1 + K_1)W_{2c} - (K_1 + K_2)W_{2c}^2 + K_2W_{2c}^3$$

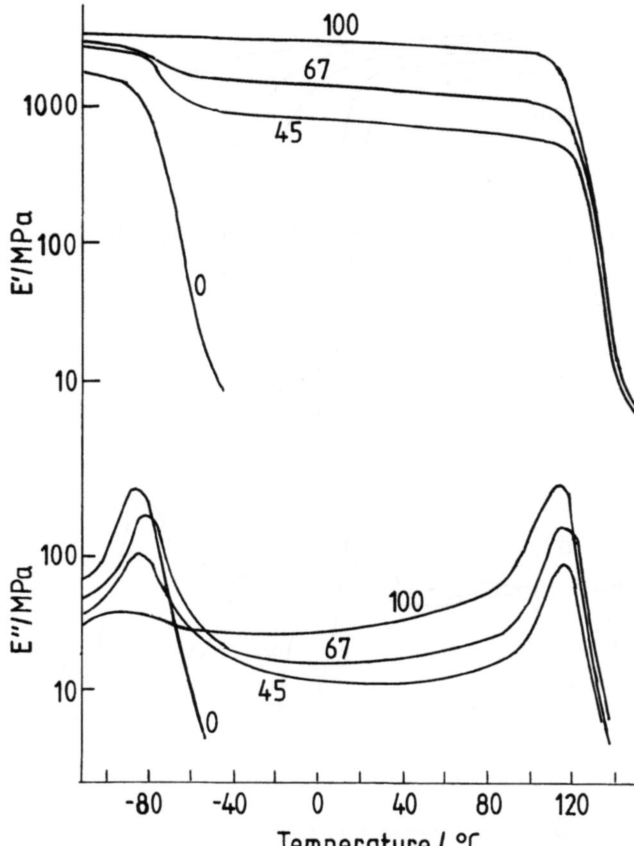

Figure 6.16 The mechanical behaviour, as a function of temperature, of a molecularly immiscible polymer mixture. The numbers on the curves are the percentages of the glassy component present.

where

$$W_{2c} = \frac{KW_2}{W_1 + KW_2} \quad \text{and} \quad K = \rho_1 T_{g1}/\rho_2 T_{g2}$$

with ρ_1 and ρ_2 the densities of the two phases.

In the Schneider expression W_1 and W_2 are the weight fractions of the softer and the stiffer components respectively, with T_{g2} being the glass transition temperature of the latter. K_1 and K_2 are given by

$$K_1 = \frac{K_1^*}{T_{g2} - T_{g1}} \quad \text{and} \quad K_2 = \frac{K_2^*}{T_{g2} - T_{g1}}$$

K_1^* relates to the interaction energy differences between hetero- and homo-contacts while K_2^* considers energetic effects on binary contacts of the

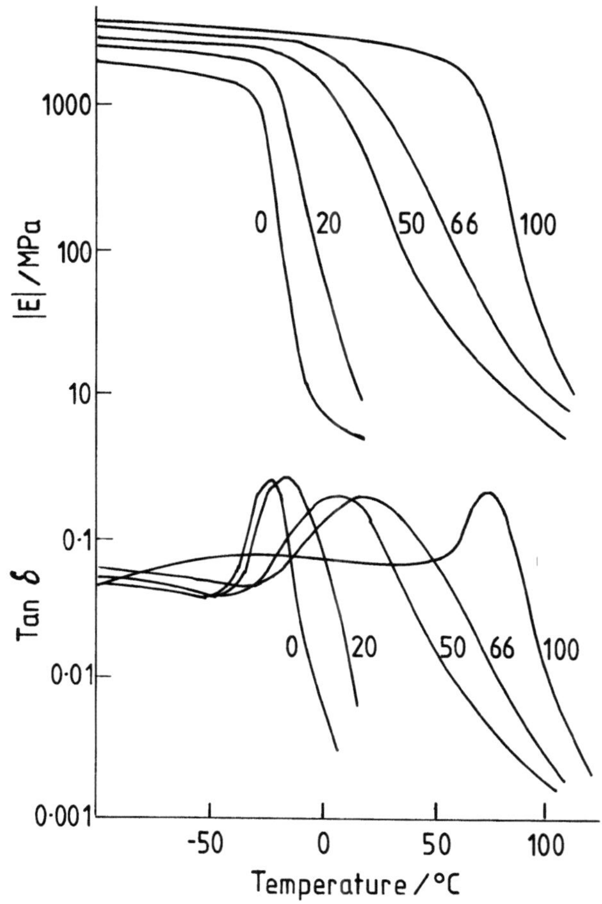

Figure 6.17 Schematic behaviour of a molecularly miscible mixture of rubbery and glassy polymers, illustrating the Gordon–Taylor equation. Numbers indicate the percentage of the glassy component.

molecular environment. The Schneider equation reduces to the Gordon–Taylor equation when $K_1 = K_2 = 0$.

Appendix

The relations between the two-suffix notation and the tensor notation are, for stiffnesses,

$$C_{aabb} = C_{ab}(a, b = 1, 2, 3)$$
$$C_{aabc} = C_{ad}(a, b, c = 1, 2, 3; d = 4, 5, 6)$$
$$C_{abcd} = C_{ef}(a, b, c, d = 1, 2, 3; e, f = 4, 5, 6)$$

and the pairs of suffixes 11, 22, 33, 23 (or 32), 31 (13), 12 (21) become, respectively, 1, 2, 3, 4, 5 and 6. Thus $C_{1111} \rightarrow C_{11}$, but $C_{1212} \rightarrow C_{66}$.

For the compliances, however, we have

$$\begin{aligned} S_{aabb} &= S_{ab}(a, b = 1, 2, 3) \\ 2S_{aabc} &= S_{ad}(a, b, c = 1, 2, 3; d = 4, 5, 6) \\ 4S_{abcd} &= S_{ef}(a, b, c, d = 1, 2, 3; e, f = 4, 5, 6) \end{aligned}$$

Stresses and strains in the tensor notation transform as follows:

$$\sigma_{11}, \sigma_{22}, \sigma_{33} \rightarrow \sigma_1, \sigma_2, \sigma_3$$

but for shear stresses

$$\sigma_{23}, \sigma_{31}, \sigma_{12} \rightarrow \sigma_4, \sigma_5, \sigma_6$$

For strains

$$e_{11}, e_{22}, e_{33} \rightarrow e_1, e_2, e_3$$

but the shears require a factor 2 in the transformation so that

$$2e_{23}, 2e_{31}, 2e_{12} \rightarrow e_4, e_5, e_6$$

If the direction cosines of the axes (1′, 2′, 3′) with respect to the set (1, 2, 3) are $l_{i'j'}$ then the tensor C_{ijkl} in the (1, 2, 3) axes becomes

$$C_{i'j'k'l'} = l_{i'i} l_{j'j} l_{k'k} l_{l'l} C_{ijkl}$$

with automatic summation over the indices from 1 to 3. $C_{i'j'k'l'}$ has, therefore, 81 components.

The direction cosines for the transformation of axes in Figure 6.14 are

$$\begin{aligned} l_{1'1} &= \cos\psi \cos\theta \cos\phi - \sin\psi \sin\phi \\ l_{1'2} &= -\cos\psi \sin\phi - \sin\psi \cos\theta \cos\phi \\ l_{1'3} &= \sin\theta \cos\phi \\ l_{2'1} &= \cos\psi \cos\theta \sin\phi + \sin\psi \cos\phi \\ l_{2'2} &= -\sin\psi \cos\theta \sin\phi + \cos\psi \cos\phi \\ l_{2'3} &= \sin\theta \sin\phi \\ l_{3'1} &= -\cos\psi \sin\theta \\ l_{3'2} &= \sin\psi \sin\theta \\ l_{3'3} &= \cos\theta \end{aligned}$$

(Note: ψ measures the rotation of the 1 axis out of the plane defined by 3 and 3′.)

Then the mean value of $C_{i'j'k'l'}$ is given by

$$\langle C_{i'j'k'l'} \rangle = \frac{1}{8\pi^2} \iiint l_{i'i} l_{j'j} l_{k'k} l_{l'l} C_{ijkl} \Phi(\theta, \phi, \psi) \sin\theta \, d\theta \, d\phi \, d\psi$$

with an analogous expression for $\langle S_{i'j'k'l'} \rangle$. The two-suffix symbols may then be derived from the tensors given above, using the symmetry of the problem.

References

1. Treloar, L.R.G. (1975) *The Physics of Rubber Elasticity*, Clarendon Press, Oxford.
2. Ferry, J.D. (1980) *Viscoelastic Properties of Polymers*, Wiley, New York, Chapter 11.
3. McCrum, N.G., Read, R.E. and Williams, G. (1967) *Anelastic and Dielectric Effects in Polymeric Solids*, Wiley, New York.
4. Nye, J. (1985) *Physical Properties of Crystals*, 2nd edn., Oxford University Press.
5. Hearmon, R.F.S. (1961) *An Introduction to Applied Anisotropic Elasticity*, Clarendon Press, Oxford.
6. Hill, R. (1963) *J. Mech. Phys. Solids* **11**, 357.
7. Walpole, L. (1966) *J. Mech. Phys. Solids* **14**, 151.
8. Takayanagi, M. (1963) *Mem. Fac. Engn. Kyushu Univ.* **23**, 1 and 41.
9. Ward, I.M. (1983) *Mechanical Properties of Solid Polymers*, 2nd edn., Wiley, New York, Chapter 10.
10. Walpole, L. (1969) *J. Mech. Phys. Solids* **17**, 235.
11. Van Fo Fy, G.K. and Savin, G.N. (1965) *Mekh. Polim.* **1**, 151.
12. Bruggeman, D.A. (1937) *Ann. Phys. Lpz.* **29**, 160.
13. Allen, P., Arridge, R.G.C., Ehtaiatkar, F., and Folkes, M.J. (1991) *J. Phys. D: Appl. Phys.* **24**, 1381.
14. Sangani, A.S. and Lu, W. (1987) *J. Mech. Phys. Solids* **38**, 1.
15. Arridge, R.G.C. (1992) *Proc. Roy. Soc.* **A438**, 291.
16. Folkes, M.J. and Ward, I.M. (1975) in *Structure and Properties of Oriented Polymers*, Ward, I.M., ed. Applied Science Publishers, London, Chapter 6.
17. Paul, D.R. and Newman, S. (eds.) (1978) *Polymer Blends*, Vol. 1 (Ch. 8) and Vol. 2 (Ch. 11), Academic Press, New York.
18. Koklas, S.N. and Kalfoglou, N.K. (1992) *Polymer*, **33**, 75.
19. diMarzio, E.A. and Gibbs, J.H. (1959) *J. Polymer Sci.* **50**, 121.
20. Schneider, H.A. (1988) *Macromol. Chem.* **189**, 1941.

Further reading

Aggarwal, S.L. (ed.) (1970) *Block Polymers*, Plenum Press, New York.

Arridge, R.G.C. and Folkes, M.J. (1972) *J. Phys. D: Appl. Phys.* **5**, 344.

Arridge, R.G.C. (1985) *Introduction to Polymer Mechanics*, Taylor and Francis, London.

Brekner, M.J., Schneider, H.A. and Cantow, H.J. (1988) *Macromol. Chem.* **189**, 2085.

Folkes, M.J. (ed.) (1985) *Processing, Structure and Properties of Block Copolymers*, Elsevier Applied Science Publishers, London.

Gordon, M. and Taylor, J.S. (1952) *J. Appl. Chem.* **2**, 493.

Hashin, Z. and Shtrikman, S. (1963) *J. Mech. Phys. Solids* **11**, 127.

Hseih, D.-T. and Peiffer, D.G. (1992) *Polymer* **33**, 1210.

Manson, J.A. and Sperling, L.H. (1976), *Polymer Blends and Composites*, Plenum Press, New York.

Olabisi, O., Robeson, L.M. and Shaw, M.T. (1976) *Polymer–polymer Miscibility*, Academic Press, New York.

7 Toughened polymers

W.H. LEE

7.1 Introduction

Commodity plastics such as polyethylene (PE), polypropylene (PP), polystyrene (PS), and polyvinylchloride (PVC) make up a large of proportion of the total tonnage of plastic currently being used mainly for non-loadbearing applications (e.g. consumer products). However, with the ever-increasing use of plastics in the areas dominated by the use of metal or ceramics, e.g. in the automobile industry, new engineering plastics, both thermoplastics and thermoset resins, have been developed which provide the combinations of lightness and good balance of stiffness, and some also in toughness, over a wide range of temperature applications.

While the development of new polymers with novel primary chain structures has continued, in parallel established materials are repeatedly modified and improved in the form of alloys or blends, largely by the innovation in synthesis and compounding. The blends and alloys of these plastics and rubbers constitute one of the most rapidly evolving areas of engineering plastics.

For ease of discussion, the blends or modified polymers discussed here refer mainly to heterogeneous (two-phase) polymers but do not include modified elastomers and glass or carbon-fibre reinforced polymers. The intention of this chapter is to highlight some of the progress which has been made on toughened polymers and their blends. The areas of work described in this chapter are of current interest.

7.2 Toughened thermoplastics

Over the last decade, there has been some progress in the development of novel thermoplastics, especially in the area of engineering plastics. An example of this is the aromatic group of thermoplastics, e.g. polyetherether ketone (PEEK), polyarylsulfone (PAS) and polyetherketoneketone (PEKK), which is used in high temperature applications. These polymers can be represented with a general formula as $(-Ar_1 - X - Ar_2 - Y -)$. Their physical and mechanical properties are determined by the choice of the X, Y and the aromatic Ar_1 and Ar_2 groups, and their distribution along the molecular chain

Table 7.1 Structures and properties relationships of some engineering thermoplastics.

Structure (repeat unit)*	Notation	T_g (°C)	T_m (°C)	G_{1C} (J/m^2)
$\{-\phi-O-\}$	PPE	80	298	—
$\{-\phi-S-\}$	PPS	90	295	—
$\{-O-\phi-SO_2-\}$	PES	225	—	—
$\{-\phi-CO-\phi-O-\}$	PEK	165	365	>4000
$\{-\phi-CO-\phi-O-\phi-O-\}$	PEEK	143	343	>4000
$\{-O-\phi-\phi-O-CO-\phi-O-CO-\phi-CO-\}$	LCP	100	331	1245

*ϕ: aromatic ring

as shown in Table 7.1. These aromatic thermoplastics are often semi-crystalline with high glass transition temperature (T_g) and melting point (T_m). They are, therefore, resistant to deformation at high temperatures but have high toughness (G_{1c}) because of the absence of crosslinking primary valence bonds, which allows the polymer chain to slip and slide during the application of stresses.

The majority of these new high temperature aromatic thermoplastics are finding application in the electrical, electronic and automobile industries while some are used as a matrix resin in continuous carbon-fibre composites for aerospace applications. Because they have inherent toughness, attempts to improve their toughness by addition of tougheners is seldom reported, except in blending and alloying with other commodity and engineering plastics.

A large number of toughened thermoplastic blends are available commercially and their properties and applications are well reviewed in the literature.

7.2.1 *Blending of thermoplastic polymers*

Improving mechanical properties such as toughness is usually the main reason for the development of novel thermoplastic alloys and blends. Other reasons for blending two or more polymers together include: (i) to improve the polymer's processability, especially for the high temperature polyaromatic thermoplastics, (ii) to enhance the physical and mechanical properties of the blend, making them more desirable than those of the individual polymers in the blend and (iii) to meet the market force. An example of the last reason is the current growing interest in the plastics recycling process where blending technology may be the means of deriving desirable properties from recycled products.

7.2.1.1 Method of thermoplastic polymer blending. The process of blending allows the intensive transfer of polymer chains occurring at the polymer–

polymer interfaces to achieve a homogeneous blend. The level of homogeneity obtained depends on the nature of the components to be blended and the blending technique employed.

1. *Mechanical blending*: Normally only produces a very coarse dispersion blend. The properties of the blends are strongly influenced by the speed and temperature of mixing. Homogeneous mixing of blends is only achieved after the melt processing stage.
2. *Solution blending*: Requires that the polymers to be blended can be dissolved in a common solvent(s). Good molecular level of mixing can be obtained with cost depending on the solvent(s) and its recovery.
3. *Polymerisation*: Mainly emulsion polymerisation. The polymers are required to be in latex or emulsion form. The mixing process of these microsize latexes and the subsequent removal of water produces excellent dispersion and distribution of discrete phase.
4. *Reactive blending*: This innovative method facilitates the ease in generalisation of new materials from highly incompatible pairs. The process often involves addition of a third reactive ingredient, usually a multifunctional copolymer or trans-reactive catalyst. Improved compatibility of reactive blends is usually attributed to the emulsifying effects of interchain block or graft copolymers that are formed during melt blending. A more homogeneous blending with high productivity can be obtained with this method but with the penalty of involving a more stringent production control.

7.2.1.2 Miscibility of blends. It is important to realise that mixing of two polymers to a homogeneous level in most cases is quite difficult. Depending on their miscibility and compatibility, some polymer pairs are almost impossible to mix with or disperse into one another. Many attempts only result in coarse aggregates or agglomeration with little or no cohesion between the two phases.

It is commonly thought that the exothermic heat generated during mixing involves specific intermolecular interaction such as hydrogen bonding, n-π complex formation and ionic interactions. However, a much overlooked contribution from intramolecular repulsion interaction to the overall heat of mixing has been recently highlighted by Paul [1] to explain the observed favourable mixing behaviour of random copolymers with a homopolymer which the corresponding homopolymers do not exhibit.

A blend which consists of two totally miscible polymers can usually be characterised by a single glass transition temperature (T_g) and homogeneous microstructures with phase size down to 5–10 nm [2]. Favourable physical and mechanical properties can be derived from the blend of two polymers which are miscible with one another. The properties of the blend are usually between those of its constituents.

With a few exceptions like poly(phenylene-ether)/PS or poly(phenylene oxide)/PS blends, where both sets of polymers in the blends are totally miscible with each other, most polymer pairs are either immiscible or almost compatible (partially miscible). High interfacial tension and disparity between the polarities of these polymer pairs resulting in a sharp interface usually exist between the phases with either very low or no interfacial adhesion.

However, many successful commercial toughened blends are either immiscible or compatible (partially miscible) and have two separate T_g and heterogeneous microstructures with dispersed phase size in micrometres as compared to nonometres for the homogeneous blends. The overall physicomechanics of these blends depends heavily on the interfacial adhesion across the phase boundaries of the two polymers. To improve this interfacial adhesion, especially for the immiscible blends, compatibilisers are added to the blends.

Compatibilisers are usually in the form of block or graft copolymers. They may be added separately or formed during compounding, mastication or polymerisation of a polymer monomer in the presence of another polymer. The copolymer compatibilisers often contain segments which are either chemically similar to those in blend components (non-reactive compatibiliser) or miscible or adhered to one of the components in the blend (reactive compatibiliser). In the case of reactive copolymer compatibilisers, the segments of the copolymer are capable of forming strong bonds (covalent or ionic) with at least one of the components in the blend. In the non-reactive copolymer compatibiliser, the segments of the copolymer are miscible with each of the blend components.

The behaviour of small amounts of copolymer–compatibiliser in an immiscible blend has been described as a classic emulsifying agent, similar to the soap molecules at an oil–water interface [3]. The emulsifying effect of compatibilisers was investigated by Noolandi and Hong [3] for immiscible blends and by Leibler [4] for nearly compatible blends.

Noolandi and Hong [3] found that the reduction of interfacial tension of copolymer in an immiscible blend is the result of the surfactant activity of the block copolymer chains. This reduction of interfacial tension increases with increasing copolymer concentration and molecular weight, causing a decrease in the interaction energy of the copolymer at the interface. They also reported that beyond a critical copolymer concentration, the surface tension remained constant.

Liebler [4] studied the interfacial properties of a nearly compatible blend, A and B, and a copolymer AB. He found that the copolymer chain improves compatibility because it has a tendency to locate itself at the interface. The presence of copolymer molecules dissolved in the main homopolymer matrices improves miscibility.

Since Paul and his co-workers [5, 6] and Heiken *et al.* [7, 8] independently showed the incorporation of these compatibilisers to improve miscibility and mechanical properties of immiscible blends in the 1970s, there has been significant effort in the development and commercialisation of compatibilisers.

7.2.2 Block and graft copolymer compatibilisers in thermoplastic blends

The success of the use of block and graft copolymers (and also random copolymer [1, 9]) as compatibilisers accounts for some of the large number of commercially available blends, e.g. HIPS and ABS. The recent increase in the interest in the reactive melt processing such as reactive extrusion (REX) and reactive injection moulding (RIM) also accounts for the increase in the level of development of copolymer compatibilisers.

Keskula and Paul [10], using methyl-methacrylate (MMA) and methyl-methacrylate-styrene (MMAS) graft rubbers (with low free polymethyl-methacrylate (PMMA) content) to blend with ABS, found an improvement in fracture toughness in MMA-graft rubber and a reduction in toughness in MMAS-graft rubber. The reduction was observed to increase with increasing styrene content of the MMAS-graft rubber. They also found that the MMA-graft rubber can be used to toughen polystyrene (PS) and HIPS, and the poor mechanical properties of HIPS/ABS immiscible blend could be improved with the incorporation of the MMA-graft rubber.

Inoue *et al.* [11] investigated the mechanical properties of a wide range of blends, i.e. PMMA/PC, PEEK/PS, PC/PPS, PC/PMMA, PC/PS and PA6/SAN, all at 70:30 ratio. They found the impact strength was much lower than for their pure homopolymer counterparts. They also found, that by adding polystyrene-co-maleic anhydride (SMA) in PA6/SAN, a significant improvement in toughness was obtained. This, they proposed, was due to the formation of PA6-SMA-graft copolymer which resulted in a fine dispersion of SAN particles.

Further work by Choudhary *et al.* [12, 13] showed that the toughness of PP/HDPE (90/10) binary blend is double that of PP but half that of HDPE homopolymer. They further found that with the inclusion of EPDM, the ternary blend of EPDM/PP/HDPE increased in toughness with increasing EPDM content up to 20%. They reported that the HDPE droplets stick to the PP matrix in the PP/HDPE blend and on addition of EPDM, both EPDM and HDPE form composite particles in the PP matrix.

Gisbergen *et al.* [14] found the EPDM particle size in the PS matrix decreases with incorporation of the diblock copolymer of PS and PB (polybutadiene). The improvement of toughness of the blends increases from 2 to 5 kJ m^{-2} for without and with copolymer respectively.

Jian *et al.* [15] designed and synthesised a PS–PC block copolymer for a PPO/PET blend. The miscibility of PPO with PS and PET with PC resulted in the reduction of the dispersed PPO phase with an improvement in toughness of the blend.

Nunes *et al.* [9] compared the use of a PMMA homopolymer, styrene-methylmethacrylate (SMAA) random copolymer and PMMA/PS blend as compatibilisers in the mixing of polyvinyl/polystyrene (PVDF/PS) blends. They found a significant improvement in the properties of PVDF/PS with PMMA as compatibiliser. With SMAA random copolymer as compatibiliser, improvement in properties decreases with styrene content. However, the styrene content of the random copolymer makes it more effective in mixing than the PS/PMMA blend.

7.2.3 Thermoplastic blends

There has been some interest in establishing the relationship between toughenability and the characteristics of modifiers [16]. Borggreve *et al.* [17] studied toughening of nylon 6 (PA6) with a range of modifiers such as ethylene–propylene–diene (EPDM), ethylene–propylene (EPM), low density polyethylene (LDPE), polyether-esters and some functional EPDM, EPR and LDPE. They reported an improvement of nylon 6 by the presence of the second phase, but the improvement appears to increase with decreasing modulus of the modifiers, with the exception of polyether-esters.

Wroctecki *et al.* [18] studied the effect of particles with an elastomeric core and a rigid PMMA shell in the toughening of PMMA. They found that the toughness increased with an increasing degree of grafting and increased molecular weight of the shell.

Gisbergen *et al.* [14] obtained a significant increase in the impact strength (up to $14\,kJ\,m^{-1}$) by subjecting their PS/EPDM blends to electron beam irradiations. They pointed out that the increment was due to the increase in grafting at the interface rather than increase in crosslinking of the EPDM rubber. Nevertheless, it has been suggested that some degree of crosslinking of the rubber modifier phase is necessary to improve the toughening effect of the rubber particles [19].

Tinker [20], in his work on PP/NR (natural rubber) blend, found substantial improvement in impact strength for slightly crosslinked NR compared to the uncrosslinked NR phase. Similarly, Dao [21] found that a slightly crosslinked EPDM rubber is more effective in toughening PP.

The effect of the presence of block or graft copolymer on the morphology of crystalline polymer blends was reported by Jang *et al.* [22]. They found that additions of EPDM, styrene–butadiene rubber (SBR), and EPR to PP resulted in less regular spherulite texture with less defined boundaries and a reduction in spherulite size in injection moulding samples. The presence of the modifiers promotes α-type as opposed to β-type spherulites.

Choudhary *et al.* [12], on the other hand, found an increase in spherulite size with addition of EPDM in PP and determined that the boundary of the spherulite is only affected at higher EPDM content ($> 20\%$). The discrepancy between Jang and Choudhary's finding could be due to the annealed and quenched thin films which were used in Choudhary's work compared to the microtomed section of injection moulding samples used by Jang. The results of Choudhary *et al.* also showed increasing toughness of PP with increasing EPDM content associated with the decrease in the degree of crystallinity.

Martuscelli *et al.* [23] studied the property–morphology relationship of EPR and EPR-graft rubber modified PA6 blends. They reported that the presence of graft rubber increases the spherulite nucleation but it also reduces the crystallisation rate and the improvement of toughness was greater than in the grafted rubber blend.

Nadkarni *et al.* [24] studied the crystallisation behaviour of different thermoplastic polyesters (PET and PBT) and a polyester copolymer with amorphous polyamide blends. They found that crystallisation of the polyester is facilitated by molten phase resulting from the blending. The crystallinity increases for PET but remains the same for PBT. They concluded that it is the inherent crystallisability of the individual polymers which determines the degree of change in both the crystallisation rate and degree of crystallinity.

Recently Teh and Rudi [25] compared melt blend of PS/LDPE and reaction extruded blends with a coupling agent. They found the reaction extruded blends method improved the impact properties within a certain level of coupling agent content. They attribute the improvement in impact to the formation of graft copolymer during reaction extrusion.

Fowler and Baker [26], using reactive blending in their work with rubber modified PS, found the improvement in impact of the blends to be a result of improved interfacial adhesion.

In the blending of high temperature engineering thermoplastics, Aghta and White [27] experimented with binary and tertiary blends of polyphenylenesulfide (PPS), polyetherimide (PEI) and polysulfone (PSF). Although the blends were immiscible, improvement of impact strength of PPS with inclusion of PEI and PSF was observed.

7.2.4 Thermoplastics/liquid crystalline polymer (LCP) blends

The development of LCP has provide a new way of generating novel blends. Unlike the polymer alloys and blends discussed in the previous sections, blending of thermotropic LCP with thermoplastics was investigated as early as in the late 1970s. The aim of blending LCP and thermoplastics, in many cases, was not just to improve toughness but to use LCP either as a processing aid or to improve the stiffness of the flexible polymers.

In using LCP as reinforcement for flexible thermoplastics, most workers related the changes in mechanical properties to the morphology of the LCP phase in the blends. This morphology is influenced by processing methods, temperature and composition.

The LCP commonly used in blends are thermotropic LCP which, because of their unique melt viscosity, can be processed with conventional extrusion, injection moulding and melt spinning equipment. A number of studies have shown that extensional flow, which is encountered during processing (e.g. injection moulding), can preferentially orient the LCP domain in the direction of flow. The oriented LCP phases can slide past one another, lubricating the polymer melt and lowering the viscosity of the blend [28].

Isayev and Modic [29] reported that injection moulded and extruded LCP/PC blends containing up to 25% LCP have spherical LCP domains dispersed in PC, while 10% LCP content produced a fibrillar LCP phase. Nobile *et al.* [30] found that a LCP/PC blend processed at 260 °C produces spherical LCP domains and at 210 °C a fibrillar LCP phase is observed.

Shin and Chung [31] reported microfibrillation of the LCP phase in the blend of LCP/PA66 as a result of extensional flow, while the spherical and ellipsoidal shaped LCP were a result of shear flow.

In nearly all the LCP/thermoplastic blends both components of the blends are immiscible, with poor interfacial adhesion resulting in loss of toughness (elongation at break). To improve the interfacial adhesion, Shin and Chung introduced long flexible aliphatic groups between the rigid main chain mesogenic groups of an LCP. The resultant fracture surface of the LCP/PA66 blend showed LCP fibres with adhering PA66 matrix which improved tensile stress while retaining toughness.

In the blends mentioned above, the shear viscosity of the LCP is lower than its thermoplastics counterparts. Isayev and Subramanian [32] recently explored blends that consists of LCP which possess higher shear viscosity than the thermoplastic (PEEK) used in the blend. They showed that at low LCP content, the impact properties of PEEK were improved and the LCP was observed to be ellipsoidal in shape and dispersed in the PEEK matrix. At a higher LCP level phase inversion occurred with the formation of PEEK fibrils in the LCP matrix.

7.2.5 Particulate modified thermoplastics

Particulate fillers including glass beads and mineral fillers such as silicate, metalflakes and talc, are generally added to improve stiffness and compressive strength, and to reduce shrinkages and the cost of the material. In most cases the addition of these fillers tends to reduce the toughness of the thermoplastic matrix, but a greater retention of the toughness can be obtained in fillers with stronger interfacial adhesion with in the matrix.

There are a few cases where toughness is increased by inclusion of fillers. At very low levels of loading (less that 4%), Bucknall and Jones [33] found an increase in toughness of PP modified with glass beads. Above this level of loading, the impact strength fell rapidly with increasing glass content.

Cotterell *et al.* [34] reported an increase in the fracture toughness of PP and ethyelene–propylene copolymer when calcium carbonate filler (omycarb) was added.

7.3 Toughened thermosets

Most of the novel thermoset resins developed over the last decade are largely used as matrices in fibre-reinforced composites for aerospace industries. As both military and commercial aircraft are pushing towards greater speed, therefore higher temperatures, these new resins are capable of withstanding elevated temperature applications via increasing rigidity and/or crosslink density of their networks. However, in almost every case the same network which provides the high temperature properties also inhibits molecular flow thus rendering the material low in toughness (Figure 7.1).

Despite their brittleness, thermoset resins are still important for use in fibre composites. One reason is that existing aircraft manufacturers possess processing and fabricating equipment which is mainly suitable for thermoset resins. The high cost of transforming them to process thermoplastics and the higher cost of thermoplastic processing itself makes the efforts to toughen existing thermoset resins much more attractive and economical. Consequently, there is a lot of interest in the toughening of these thermosets.

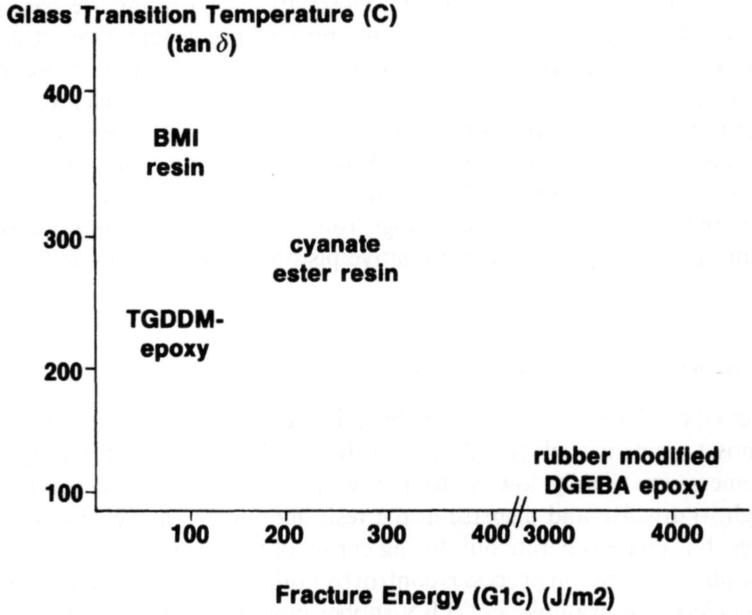

Figure 7.1 Relationships of T_g and fracture toughness of some thermoset resins.

The fracture toughness of thermosetting resins of 50–300 J m^{-2} is marginally higher than inorganic glass. However, by incorporating toughener into thermoset resins such as epoxies, the level of toughness can be brought into line with some thermoplastics (Figure 7.1).

Much of the published work concerning toughening of thermosetting resins involves incorporation of soft elastomeric substances into the resin matrix. As the high temperature requirement becomes more and more critical, incorporation of engineering thermoplastics as toughening agents has become more and more popular.

There are several thermoset resins which are currently receiving a great deal of attention in an effort to improve their toughness. Examples of these are the epoxy resins, cynate ester resins (CE) and polyimide (bismaleicimide-BMI) resins.

Modification and variants of epoxy resins have constantly been developed by resin producers over the last few decades. There is a vast number of different epoxy resins and curing agents which can be used in different combinations. The chemistry and applications are well reviewed in the literature [35].

Other resins introduced are cyanate ester (CE) resins (which include both dicyanate resins and bismaleimide-triazine (BT) resin) [36] and bismaleimide (BMI) resins [37]. These resins have similar processing characteristics to the epoxy resins, but claim to possess better mechanical properties. They were designed to fill the maximum service temperature gap between epoxy (150 °C) and polyimide (250 °C) resins.

Most of the maleic end group thermoset PI (BMI) resins are based on aromatic diamine. Some of these resins possess processing characteristics similar to those of epoxy resins. The norbornene end group PI thermoset resin most widely accepted by the aerospace industry is PMR-15. In the production of this resin the polymerisation of the monomer reactant (PMR, prepolymer molecular weight = 1500) is performed during resin processing and curing. This resin is almost exclusively used in glass and carbon fibre composites. Other applications of imide resins range from jet engine trans scouls, radomes and aircraft landing gears to automative pistons, tank bearings and missile fins.

7.3.1 Blending of thermosetting resins

The compatibility requirement of blending the elastomer modifiers with thermoset resins is slightly different to that of thermoplastic blending. The elastomers are usually low molecular weight liquids and are compatible enough to dissolve and disperse in the resin monomer (or their lacquer) on storage, but phase-separate out during curing of the resin.

The phase separation process is control by both thermodynamic and kinetic factors [38]. For example, in CTBN rubber modified epoxy, the increase in molecular weight during curing, for both the rubber and the epoxy, reduced the entropy of mixing. This, together with the positive (endothermic) enthalpy, increases the likelihood of a two-phase morphology. The overall resultant morphology is also influenced by reaction rate at a given temperature and the period over which phase separation can take place.

Conversely, in the mixing of two relatively low molecular weight components with relatively matching solubility parameters (δ) of CTBN rubber and epoxy, the negative entropy contribution dominates the free energy of mixing. In blends of thermoplastics with thermosets, the significance of entropy factors is greatly diminished. Consequently the miscibility is influenced by the difference in the δ values [39].

Most of the thermoplastics used in modified thermoset resins are usually high temperature engineering polymers, such as PES, PI and PAS. These modifiers are either incorporated by hot melt blending or by solution blending. However, the same solvent which is used in the blending process often turns out to be the solvent which will affect the cured laminate properties unless the modifiers contain reactive end groups which crosslink with the matrix network. Modifiers which can only be blended in by the hot melt process are either dissolved or dispersed in the resin mixtures. For fibre composite applications, the particle size distribution of the finely dispersed modifier must be below 7–8 μm, which is equivalent to the average fibre spacing in a fibre composite.

In general, a small degree of chemical reaction between the resin and the modifier is often desirable for effective toughening. The blending methods described above for epoxy resins are also applicable to other thermosetting resins.

7.3.2 Toughening of thermoset resins with soft inclusions

The current use of elastomeric particles in toughening of thermosets is mainly for low temperature application (below 100 °C). The majority of these are toughened with CTBN rubber and coverage of the work carried out on these compounds can be found in the literature [40–52].

A new elastomeric toughener was recently reported by Pocius [52] using 'phase separated' acrylic elastomers suspended in uncured epoxy via a mechanism known as 'steric stabilisation'. A much higher toughness (peel strength) improvement was obtained by this method than by conventional *in situ* phase separation on curing of CTBN modified epoxy. Lee *et al.* [53, 54] using PMMA–NR graft copolymers, found a substantial increase in fracture toughness for the low crosslinked piperidine cure diglycidyl bisphenol-A (DGEBA) epoxy and for the highly crosslinked diamino diphenyl sulphone (DDS) cure tetraglycidyl diamino diphenyl methane (TGDDM) epoxy.

Rubber toughening of cynate esters was studied by Yang *et al.* [55]. They compared the effect of core-shell rubber particles in cynate ester resins, DGEBA and TGDA epoxies and found the improvement of toughness to be even greater in cynate ester resin than in the relatively ductile DGEBA epoxy systems. They suggested that the toughenability of the thermosets may be related to the molecular inhomogeneity of the network structures.

Toughening of highly crosslinked BMI involves the same principle as for epoxies. Shaw and Kinloch [56], using liquid CTBN rubber in an eutactic mixture of BMI (Compimide 353A) resin, found that the resultant two-phase morphology increases the fracture toughness. Stanzenberger *et al.* [57] obtained similar results but reported a substantial reduction in T_g with the incorporation of rubber in the BMI resins.

7.3.3 Toughening of thermoset resins with thermoplastic inclusions

Elastomeric toughening has proved to be successful with epoxy adhesives but not with high performance composites because of the problems of transferring the matrix toughness to its fibre composites [58, 59], the reduction of interlaminal shear stress caused by the inclusion of the low shear modulus rubber particles, and the long-term elevated temperature requirement. Alternative approaches for using thermoplastics as modifiers in epoxy resins have been explored.

The addition of thermoplastics into epoxies has been made in commercial applications for the last decade. However, the original intention was to control the flow behaviour of the epoxy-fibre prepreg during its curing cycle. The inclusion of thermoplastic particles which are believed to improve the toughness of the matrix has been the subject of a great deal of research [39, 60–66].

Bucknall and Partridge [39] reported that the phase behaviour of the PES in epoxies depends on the nature of the epoxies and the curing agents used. In their work, they found no second phase and only modest improvement of fracture toughness in PES modified TGDDM/DDS systems. However, phase separated particles of PES were observed in the triglycidyl para-amino phenol (TGPAP) resin/DDS/dicydiamide (Dicy) system.

Raghava [60], using aromatic anhydride as a curing agent in PES modified TGDDM epoxy, found a two phase morphology with modest improvement in fracture toughness. He also found that with low molecular weight PES, the separated PES phase is in the form of agglomerate while with high molecular weight PES, the PES phase is spherical in shape [61].

Bucknall and Gilbert [62] experimented with PI (Ultem 1000) as a toughener in TGDDM/DDS epoxy. They found that the dispersed PI particles increase the fracture toughness relative to the PI content despite the change in morphology after 20% PI loading when phase inversion occurred.

Hourston and Lane [63] used similar PI in toughening TGPAP/DDS systems. They found dispersed PI particles in samples with less than 15% PI content and noted that fracture toughness increases with PI content until phase inversion occurs. They also reported that the toughness is further increased with post-curing.

Toughening of low crosslinked epoxies (low glass transition temperature T_g) with thermoplastics was investigated by Hedrick [64]. In combinations of hydroxy terminated poly(arylethersulphone) with a difunctional DGEBA epoxy system, he found a two-phase morphology with two-fold improvement of toughness. Yamanaka and Inoue [65], using PES in a DGEBA/DDS system, again found two-phase morphology with an increase in toughness (peel strength).

Cecera and McGrath [66], using the same DGEBA/DDS systems but with amine-terminated polysulfone (PSF), found that the increase in toughness in this two-phase system is strongly influenced by the molecular weight of PSF. With PEK as the toughener, they found gross incompatibility between the epoxy and the PEK, especially with very low molecular weight PEK, and this is reflected by the modest increase in toughness observed.

Toughening of cyanate ester resins with different thermoplastics has been reported by Shimp *et al.* [67]. The relatively ductile nature of these resins, compared to TGDDM epoxy for example, resulted in a greater increase in toughness when incorporated with thermoplastic tougheners. Two-phase morphology was observed in most cases and phase inversion occurred at high toughener contents.

The use of thermoplastics in toughening of BMI resins is seldom reported. However, Figure 7.2 shows a two-phase morphology of 15% PI (Ultem 1000) in BMI resin (Compimide 353A/Xu 292B). In this case, the resin was hot melt blended at 110 °C with a curing temperature of 180 °C for 6 hours, and post-cured at 240 °C for 6 hours. The resultant fracture toughness (G_{1c}) increased from 170 J m^{-1} for unmodified to 210 J m^{-1} for the PI modified BMI resin without any reduction in glass transition temperature.

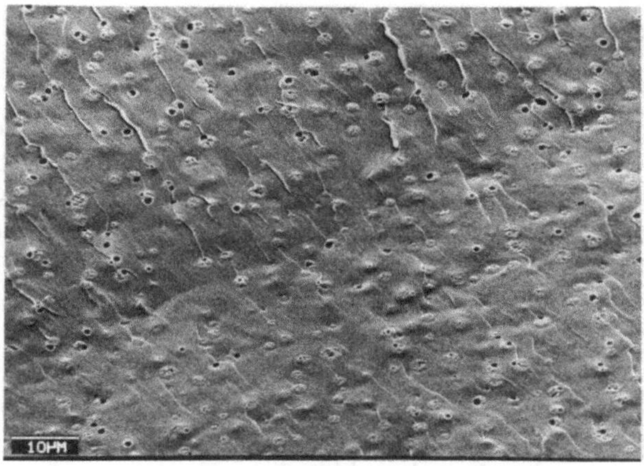

Figure 7.2 SEM micrograph of PI modified BMI (C353/Xu292B) matrix fracture surface showing the presence of dispersed PI phase.

Most of the toughening work on BMI involves flexibilised co-reactants such as bis(allyl phenyl) or bis(propenyl phenoxy) compounds. The copolymerisations and chain extension reactions lead to tougher networks but also, in most cases, a reduction in glass transition temperature.

7.3.4 Toughening of thermoset resins with rigid inclusions

Most of the published work on toughening of thermosets with rigid inclusion (particulate fillers), such as glass beads or other fillers, use epoxies as the matrices. As with thermoplastics, particulate fillers are added to reduce the cost of materials, and to control shrinkages and thermal expansion. The addition of particulate fillers usually lowers the stress at which fracture occurs in compression and improves the Young's modulus. Unlike in thermoplastics, the inclusion of these fillers increases the fracture toughness of the matrix.

Moloney *et al.* [68] reported that the increment of the fracture toughness (K_{1c}) of the silica-filled epoxy is not affected by the surface treatment of the fillers or the filler size. Spanoudakis and Young [69] demonstrated that higher K_{1c} values were found in systems with poor particle/matrix adhesion but with greater reduction in Young's modulus.

Kinloch *et al.* [70] showed that addition of glass spheres to rubber modified DGEBA resins, forming hybrid particulates, further increases the fracture toughness of the rubber modified epoxy. The increment was observed to be higher for surface treated glass spheres than for the untreated ones. Their result was confirmed by Low *et al.* [71], who used zirconia and alumina fibres as the filler in rubber modified hybrids and Moloney *et al.* [68] who used hollow glass spheres in epoxies.

7.3.5 Semi-interpenetrating network

Another approach to toughening a thermosetting polymer is via the formation of interpenetrating networks (IPNs). This can be described as a blend of two or more polymers which are crosslinked in such a way that the networks are physically interlocked but not chemically bonded. In the case of semi-interpenetrating networks (SIPNs), only one polymer is crosslinked while the other polymer is linear or branched, usually a thermoset resin and a thermoplastic. The main reason for forming an SIPN is to combine the best physical and mechanical properties from each material.

Galbriath *et al.* [72] reported that oligomer methacrylatemethylmethamethacrylate copolymer (ICI Modar resins) can be toughened by forming SIPNs with polyurethane. In SIPNs with good phase separation a significant improvement in toughness can be achieved.

Hartness [73] mixed a copolyester-carbonate thermoplastic with dicyanate bisphenol A resin (cyanate ester resin) to form an SIPN with high temperature stability and toughness improvement. The SIPN morphology showed two discrete phases.

Sefton *et al.* [74] reported a spinoidal morphology in the SIPN of thermoplastics and epoxy blends gave the best toughness improvement.

St. Clair and Hanky [75] studied the formation of SIPNs of a tough but difficult to process condensation thermoplastic PI (NR150-B2) and an easy to process but brittle addition thermosetting PI (PMR15). Besides overcoming the processing difficulty of thermoplastic PI, the toughness of the thermoset PI was significantly improved while a strength retention up to a temperature of over 300 °C was observed.

Recently Pater [76] reported using a similar combination but with a linear Larc-RP41 thermoplastic PI. He obtained similar improvement in toughness from the resulting blend but with some minor sacrifice of high temperature properties. The thermoplastic PI in this case behaves like a plasticiser rather than a toughener.

7.4 Toughening mechanisms

Most thermoset and a few thermoplastic polymers fail in a brittle manner when subjected to tensile loading, but the same polymers will exhibit ductile behaviour under compression, even the highly crosslinked TGDDM/DDS epoxy system [77]. The inherent ductility of a polymer matrix depends on its ability to undergo plastic deformation during stress. This is influenced by test conditions and geometry. Incorporation of tougheners can increase the fracture resistance of a polymer either by enabling plastic deformation to occur in the matrix, increasing the plastic deformation which is occurring in the matrix or undergoing fracture themselves. Depending on the nature of the polymers and the tougheners, different toughening mechanisms may be operative in different toughened materials.

7.4.1 Toughening mechanisms in thermoplastics

It is well established that rubber particles with low moduli act as stress concentrators in both thermoplastics and thermoset resins, enhancing shear yielding and/or crazing, depending on the nature of the matrix.

Shear yielding in a polymer matrix involves macroscopic drawing of material without a change in volume, and the process takes place at an angle of about 45° to the tensile axis. For shear yielding to occur the presence of a shear stress is required as given by the Von Mises criterion [78]. Depending on the polymer, the yielding may be localised into shear bands or diffused throughout the stress region. It is initiated by a region of high stress concentration due to flaws or polymeric inclusions such as rubber particles.

Crazing, on the other hand, is a more localised form of yielding and occurs in planes normal to the tensile stress. These are voids at the crack tip interconnected by highly drawn fibrils which also form the craze walls and contain approximately 50% voids ranging in size from 20 Å to 200 Å [79]. Crazes grow by microscopic internal drawing of materials from the craze walls so as to increase the fibril length. They nucleate at the interface of high stress concentration caused by inhomogeneity such as flaws or polymer inclusions.

The theory of multiple crazing in rubber toughened thermoplastic was proposed by Bucknall and Smith [80]. According to the theory, crazes are initiated at points of maximum principal strain, which are usually near the equator of the rubber particles, and then propagate outwards along the plane of maximum principal strain. The process is terminated when the stress concentration falls below the critical level for propagation, or when a large particle or other obstacle is encountered. Thus the rubber particles are able to control craze growth by initiating and terminating crazes (Figure 7.3).

Bucknall *et al.* [81] later proposed that shear yielding and massive crazing are the two energy absorbing mechanisms in rubber modified plastics. According-

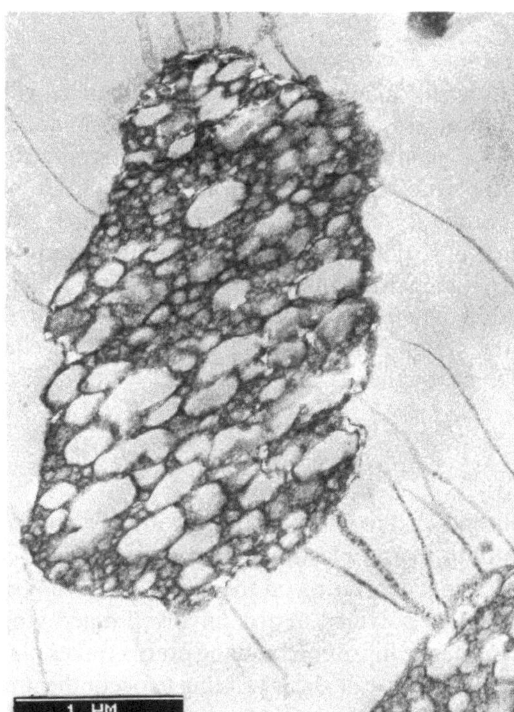

Figure 7.3 TEM micrograph taken from a tensile fractured HIPS specimen showing crazes initiating and terminating at the surface of the rubber particles.

ing to them, areas of stress concentration produced by the rubber particles are initiation sites for shear band formation as well as for crazes, and that shear yielding is not simply an additional deformation mechanism, but an integral part of the toughening mechanism. Since the molecular orientation of shear band is normal to craze, shear bands are able to limit the growth of craze. Therefore, as the number of shear bands increases, the length of newly formed crazes will decrease. Consequently a large number of short crazes is generated.

Beside limiting craze growth, another possible interaction by shear bands is the arrest of a growing craze as a result of a shear band nucleated at the craze tip (the formation of shear bands relieves the hydrostatic stresses which are required for craze growth) [82]. Also, when a craze intersects a pre-existing shear band, the craze growth continues inside the shear band at a different path to its normal trajectory, and re-emerges into the underformed matrix [82, 83]. The existing shear band can also act as a preferential site for craze nucleation [84]. In polymers which form crazes and shear bands readily, the interaction between the two will increase the overall toughness of the material.

The process of ductile fracture which occurs in material containing voids has also been observed in metals. Theoretical model calculations have shown that these voids reduce the constraint of shear stress to yield and that growth of the voids promotes shear band formation in the plastic yield zone between the voids [85–89].

The process of voiding as a toughening mechanism has been suggested by some workers [90–94]. To date, there is very little published work on the fracture properties of thermoplastic polymers containing voids. The ability of voids to lower the hydrostatic tension at the crack tip was suggested by Hobbs [90] who found that the fracture toughness of glass-filled polycarbonate foam increase with increasing void content. The mechanism suggested was that the presence of voids reduces the hydrostatic constraints and allows some shear yielding to occur in the cell wall.

The occurrence of matrix voiding as a mechanism of toughening was proposed by Yee *et al.* [91–93]. They suggested that in some thermoplastics which failed predominantly by shear yielding, the local build-up of hydrostatic stress at the crack tip is relieved by the process of voiding. The initiation of voids and their subsequent growth enhance the matrix flow. Yee *et al.* pointed out that toughening by voiding occurs most readily in polymers whose matrices are ductile but notch-sensitive like PC. Recent work by Borggreve *et al.* [94] found similar toughening mechanisms for the rubber modified nylon (PA6) blend.

7.4.2 Toughening mechanisms in thermoset resins

Toughening mechanisms in thermosetting resins, especially epoxies, have been the subject of quite intensive studies by numerous researchers over the last few years. Because of the range of crosslinked densities, and therefore of matrix ductility, that can be produced with a single resin system depending on the tougheners, there is no one generalised mechanism that can be used to describe the toughening of epoxies or other thermoset resins. In most cases toughening mechanisms of thermoset resins are combinations of a number of processes which are outlined in the following sections.

7.4.2.1 Tearing and ductile drawing of dispersed phase. Rubber tearing as a toughening mechanism was proposed by Kunz and Beaumont [46] to account for the toughening effect observed in their rubber modified epoxies. This quantitative theory attributes the toughening solely to the rubber particles.

According to the theory, a crack in a rubber toughened epoxy propagates through the brittle epoxy matrix and leaves the particles bridging the crack as it opens. The energy required to stretch and tear the particles accounts for the increment in the fracture toughness values. According to the model, the elastic strain energy dissipation during rubber tearing is expressed as:

$$\Delta G_{1c} = \left(1 - \frac{6}{\lambda_t^2 + \lambda_t + 4}\right) 4\Gamma V_p$$

where ΔG_{1c} is the increase in G_{1c}, λ_t is the tear strain at failure or the stretched particles, Γ is the tearing energy of the particles and V_p is the volume fraction of the particles.

This deformation is also applicable to those thermoplastic-toughened thermoset resins in which the particles are well-bonded to the matrices. In such cases, extensive drawing of these particles results in a significant improvement in toughening when compared to rubber tearing. This theory has found good agreement in the rubber [51] and PI [62] toughening of TGDDM/DDS epoxy systems where negligible plastic yielding has occurred.

Bucknall and Gilbert [62] described the same toughening in the PI modified TGDDM/DDS epoxy but instead of tearing of rubber, the ductile drawing of thermoplastic PI particles was proposed.

In other toughened systems where considerable plastic deformation is observed, coupled with a large improvement in fracture toughness, other toughening mechanisms must be considered.

7.4.2.2 Crack tip blunting of resin matrices. The recent toughening theories in rubber modified epoxies proposed by Bascom *et al.* [47], Kinloch *et al.* [48] and Yee and Pearson [49] are all based on similar mechanisms which can be summarised as follows.

Hydrostatic tensions are generated in well-bonded rubber particles during curing and then cooling of rubber modified polymers. During loading of a specimen, the triaxial stress state generated ahead of the crack tip, together with the hydrostatic tensions in the rubber particles causes the rubber particles to cavitate either in the particles or at the particle/matrix interface. (The ability of rubber to cavitate under high triaxial stress has been shown by Gent [95].)

This cavitation or voiding process relieves the build-up of triaxial at the crack tip.

The presence of rubber particles produces stress concentrations at their equators and also provides sites for the formation of shear bands (rubber particles able to initiate shear deformation by producing a local increase in the octahedral shear stress).

The cavitation/voiding process of rubber particles increases the stress concentration effect of the particles, this will not only relieve the hydrostatic tension at the crack tip, it also allows the shear band to grow by further reducing the octahedral shear stress to yield. Consequently, further growth of voids occurs which in turn promotes further yielding. The whole process creates a plastic yield zone containing distorted and voided particles as shown in Figure 7.4. Such zones produce a crack tip blunting effect releasing the stress concentration at the crack tip and delaying crack propagations. The crack will eventually propagate through the voided plane.

Toughening by voiding has recently been demonstrated by Huang and Kinloch [96]. By the inclusion of 17% void in a DGEBA epoxy system, they found a significant increase in fracture toughness, and the fracture surface resembled those of rubber modified epoxies.

Evans *et al.* [97] recently have proposed a quantitative toughness model which includes both rubber tearing behind the advancing crack tip and plastic deformation zone ahead of the crack tip.

7.4.2.3 Crack pinning mechanism of impenetrable and semi-penetrable particles. The increase in toughness of impenetrable particulate modified brittle polymers has been explained by the crack pinning mechanism which was first proposed by Lange [98] for a two-phase brittle material.

According to the theory, when a crack with a crack front of unit length propagates in a stressed solid meets an array of well-bonded solid particles, the crack front bows out between the particles while still remaining pinned at all positions where it encounters the particles, forming a secondary crack. During this initial stage, a new surface is formed and the length of crack front is increased due to the change in shape. Excess energy is then required to create a new fracture surface and to supply the newly found non-linear crack front, which is assumed to possess line energy. Lange gave the relationship for the increase in the fracture energy as:

$$\Delta G_{1c} = T/2b$$

where T is the line energy of the crack front and b is the interparticle spacing.

Lange's theory was latter extended and modified by Evans [99] and Green [100].

The crack pinning toughening mechanism for impenetrable particulate filled thermosets has been used with great success by Moloney *et al.* [68]. They found the mechanism accounted for the observed increase in toughness of their glass bead filled epoxy systems.

A combination of crack pinning and crack tip blunting was proposed by Spanoudakis and Young [69], Kinloch *et al.* [70] and Low *et al.* [71]. Much of the work reported found crack tip blunting is the dominant mechanism of toughening in epoxy systems with rubber modified particulate hybrids and/or testing at elevated temperatures. At low testing temperature or with only particulate filled (without rubber inclusion), toughening by crack pinning was observed.

The theory of crack pinning originally applied to impenetrable particles which pin the crack front but do not themselves fracture under the stresses in the systems. However Green *et al.* [100] have extended and modified the theory to account for relatively tough particles. In this case, the strain transferred by the matrix crack on to the ductile particles can be accommodated by the plastic deformation of the particles until it has circumvented by the crack.

This mechanism was suggested by Lee *et al.* [53, 54] in study of a PMMA/NR graft copolymer modified TGDDM/DDS epoxy where micrographic features resembling those of crack pinning were observed and the mechanism accounted for the observed increase in fracture toughness.

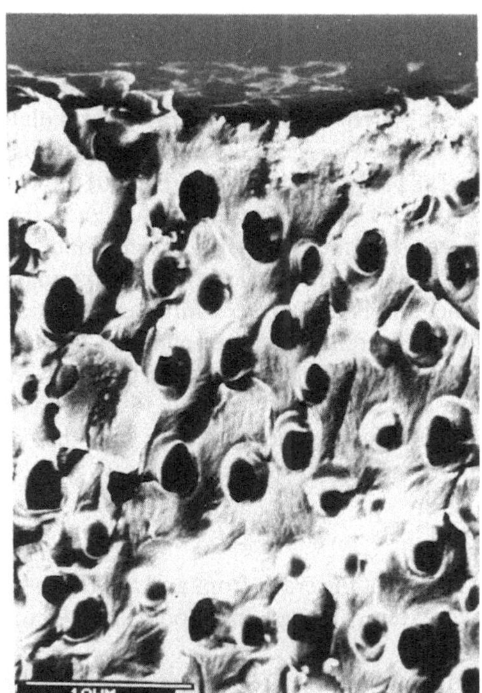

(a)

(b)

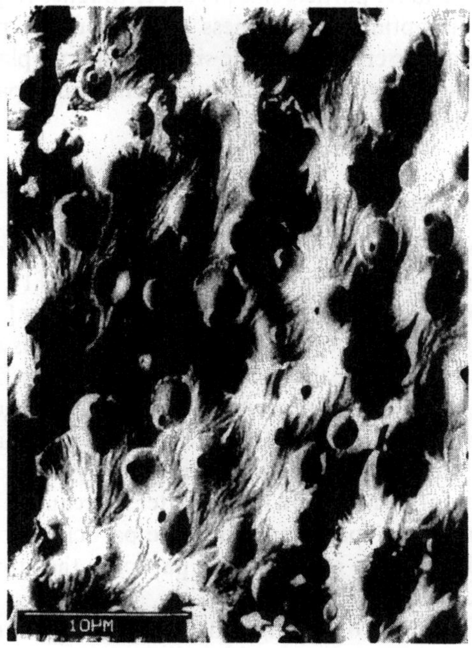

(c)

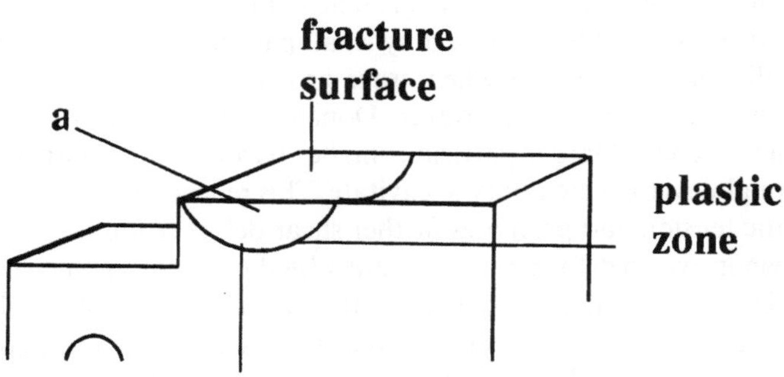

Figure 7.4 SEM micrographs of CTBN rubber modified DGEBA/pip epoxy taken from beneath the fracture plane at the region (a) beneath the voided area and (b) at the peripheral of the voided area as indicated by the accompanying schematic diagram.

7.5 Particle size distributions

The importance of rubber particle size in rubber toughened plastics cannot be over-emphasised if optimum toughness from modified polymer is to be achieved. In general, the critical particle size for a thermoplastic with a failure mechanism including crazing is much better defined compared to thermoset resins or thermoplastics which fail by shearing.

7.5.1 Particle size effect in toughened thermoplastics

In toughened thermoplastics, like HIPS for example, particle sizes of less than 1 μm in diameter do not provide effective toughening [19, 101], while PVC may be toughened by particles less than 0.1 μm in diameter [19]. This suggests that the critical particle size for each individual polymer is different.

In PS, which fail by crazing, the role of rubber particles will be to initiate and terminate crazes. It has been reported that particles larger than 1 μm are capable of controlling craze growth as a result of their ability to initiate and terminate crazes while smaller particles are not only poor craze terminators but also poor initiators. The reason for the latter is because a high stress region over several interfribrilla spacings around the particles is required for craze network to develop. In the case of small particles, the stress concentration effects are too localised [82].

In ABS, which is capable of crazing and shear yielding, the critical size which has been suggested is around 0.26–0.46 μm. It was reported in this case that particles larger than the critical size control craze growth like those observed for the HIPS, while the smaller particles promote shear deformation. Dillon and Bevis [102] suggested that the shear process in the ABS is diffuse in nature therefore it is unable to control craze growth, unlike shear bands. Consequently, shear yielding in ABS, promoted by small particles, will not contribute to the overall toughness except when craze growth is being restricted by the presence of larger particles. However, Donald and Kramer [82] concluded that in the absence of larger particles, smaller particles will contribute to the overall toughness by their ability to cavitate. This relieves the local build-up of hydrostatic tension and promotes further shear deformation.

In polymers which deform predominantly by shear yielding, such as PC, the particle size effect is believed to be minimal. The only requirement is that there should be a sharp interface between the dispersed and continuous phase. However, as shear yielding is enhanced by the cavitation of particles, the ability of a particle to void or cavitate in relation to its size is unclear at present.

7.5.2 Particle size effects in toughened thermosets

The effects of particle size in thermoset resins have also been investigated [40–44, 103, 104]. Since crazing is believed not to play a role in the toughening of thermosets [46–50], the effect of larger particles centres on their ability to cavitate under high triaxial stresses, while smaller particles promote shear yielding.

Sultant and McGarry [103] measured the biaxial yield behaviour of epoxies containing large (1–22μm) and small ($<0.1\,\mu$m) rubber particles. They found that the yield criterion could be expressed in a modified Von Mises form:

$$\tau_{oct} = \tau_0 - \mu P$$

where τ_{oct} = octahedral shear stress, τ_0 = intrinsic octahedral shear strength of the material, P = the mean hydrostatic pressure and μ = the pressure coefficient which is associated with the sensitivity of the material to the hydrostatic stress component.

Together with some optical examinations, Sultant and McGarry [103] concluded that larger particles promote microcavitation while smaller particles enhance the shear yielding process.

Bascom *et al.* [104] used a combination of liquid and solid CTBN rubbers to produce a dual particle size distribution of 0.5μm and 1–2μm respectively. They found that smaller particles deformed principally by voiding and induced local shear yielding. Larger particles produced localised yielding in the surrounding matrix which was facilitated by the presence of smaller particles. Since no large particles were observed on the fracture surfaces, it is unclear whether the deformation of the particles occurred before fracture. It must also be pointed out that the small practicle size used by Sultan and McGarry was less than 0.1 μm as compared to 0.5μm in Bascom's work.

Most other reports of the effects of particle size on toughened epoxies are from studies of the fracture surfaces and morphologies in relation to the extra toughening effect of a bimodal particle size system over those systems which contained only a single particle size [40–44].

7.5.3 Bimodal particle size distribution

From the discussion above, the effect of different particle size distributions in their ability to toughen a thermoplastic will depend very much on the matrix's failure mechanism(s). In other words, in polymers which fail by crazing only, dual particle size distribution would not further enhance the degree of toughening but for polymers which fail by crazing and shear yielding, like ABS, the presence of two different particle sizes (one above and one below the critical size diameter) would improve the toughness of the matrix compared with a polymer with single particle size.

It is interesting to note that, for both thermoplastics and thermoset resins, much of the published work with bimodal particle distributions contains occluded matrix in the large rubber particles. The amount of occluded matrix increases the modulus of the rubber particles, thus leading to a reduction in the stress concentration factors which should in turn increase the constraint on the

Table 7.2 Comparisons of mechanical properties of CTBN rubber to PMMA-NR graft copolymer (MG) toughened DGEBA epoxies [52].

Resin systems	Volume fraction of rubber V_p(%)	Yield stress σy (MPa)	Stress intensity factor K_{1c} (MN m^{-2})	Fracture energy G_{1c} (J m^{-2})
DGEBA + pip	—	63.1	1.0	400
DGEBA + pip + CTBN × 8 (10%)	12.0	57.4	2.17	3680
DGEBA + pip + MG30 (1%)	2.2	61.4	3.17	4696
DGEBA + pip + MG50 (1%)	3.15	62.2	3.45	4898

yield stress. On the other hand, the occlusion results in an increase in the effective volume fraction of the dispersed phase. This will increase the interaction between the stress fields of adjacent particles and thus decrease the yield stress constraint. Since the latter mechanism is believed to be more dominant [102], it is possible that the increase in fracture toughness is not due to bimodal distribution of particle size effects but to the increase in the volume fraction of the dispersed phase.

Recently, using PMMA-NR graft copolymer in toughening of epoxies, Lee *et al.* [53, 54] were able to obtain a much higher toughness improvement with a very small amount of graft copolymer content (and volume fraction) than with CTBN rubber at a higher content (Table 7.2). This effect was observed to be the result of the bimodal particle size distribution of the graft copolymer modified system with large particles of 0.1–3 μm and small particles of less than 0.1 μm as shown in Figure 7.5. This particle size distribution produced a stress whitening band at 45° across the gauge width (Figure 7.6) of the tensile test samples instead of diffused whitening throughout the gauge length of the mono particle size CTBN rubber modified samples.

7.6 Influence of processing on toughness of modified polymers

The methods for the processing of thermoplastic polymers are well established. There is a large amount of literature which provides information about the various processing techniques and the effects on morphology, e.g. crystallinity and orientation of the dispersed phase.

The effect of orientation of materials which affect morphology will influence the physical and mechanical properties of the materials as well as the shape of the mouldings. This is especially true for TP/LCP blends [28–32] some of which have been pointed out in section 7.2.4. With the advance of shear control injection moulding (SCORIM) and extrusion (SCOREX) [105] *in situ* fibre

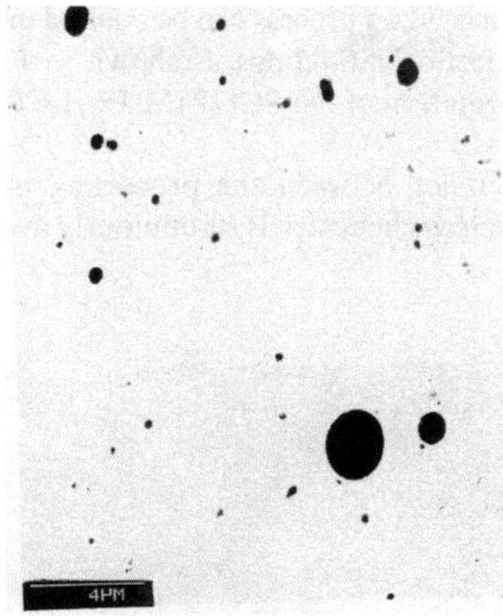

Figure 7.5 TEM micrograph of OsO_4 stained MG30 rubber particles in DGEBA epoxy showing bimodal distribution of particle size.

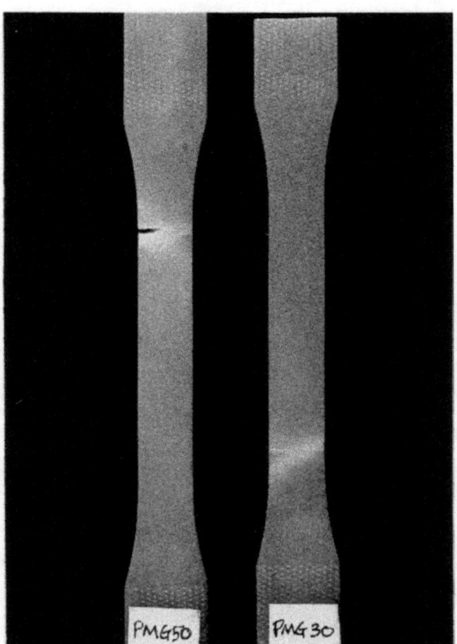

Figure 7.6 Photograph of fractured tensile test specimens of MG30 and MG50 showing whitening bands occurring at 45° to the principal stress direction.

formation during the moulding process can be realised in a greater degree than with conventional injection mouldings as shown in Figure 7.7. The higher degree of the fibril formation of the SCORIM PC/LCP mouldings increases toughness and stiffness.

The principal difference between the processing of thermoplastics and thermosets is that whereas chemistry is scrupulously avoided in the former, it

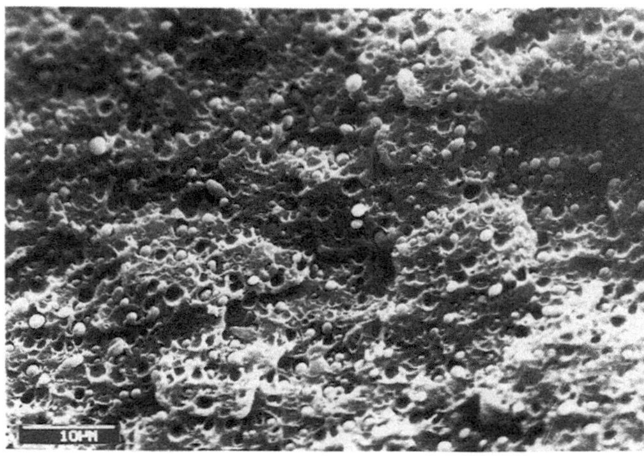

(a)

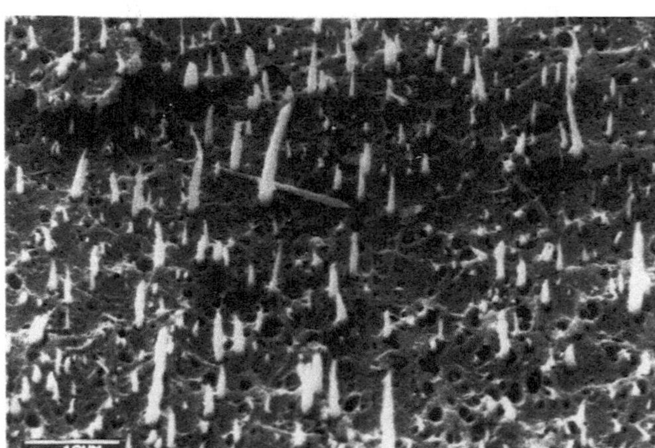

(b)

Figure 7.7 SEM micrographs of LCP/PC blend injection moulded with (a) conventional and (b) multi-live feed (SCORIM) injection moulding machine (the micrographs shown are taken from the core region of the fractured samples).

is required for the latter. In the processing techniques which provide ease of shaping, such as injection moulding and resin transfer moulding, the mechanical properties of the final product will depend heavily on the resin compositions, cure time and temperatures.

In other processing methods such as for fibre composites and some adhesives, where the material is supplied either in prepreg or preformulated resins, the methods adopted in preparing the formulated resins will also influence the properties of the cured systems.

An example of this is the preparation of a thermoplastic modified thermoset resin such as epoxy. Solution blending is used instead of hot melt since the former method is much safer and less time-consuming. This method is especially popular in the prepreg (preimpregnated fibre) where a higher tack level may be obtained. However, solution blending is only applicable to thermoplastics which can be dissolved or swelled by the solvent. Unfortunately, the two methods of mixing sometimes produce matrices with different morphologies. An example of this is shown in Figure 7.8 which is of similar formulation and cure schedule to that of Figure 7.2. However, the mixing in the example in Figure 7.8 was carried out via solution blending in dichloromethane and then transferred to the fibre from the thin film of the formulated resin. The fracture surface (with carbon fibre reinforced) does not show the presence of dispersed phase which is clearly observed in the neat resin of Figure 7.2.

One reason which may account for differences in observed morphologies is the advance on cure during mixing stage. This is especially true for epoxy mixing [106] but not for the BMI resin where very high cure temperature is required for a long period of time to initiate any reaction. The probable

Figure 7.8 SEM micrograph of G_{1c} fractured PI modified BMI (C353/XU292B) matrix in carbon fibre (T800) reinforced composite (no dispersed PI phase could be observed).

reason is the effect of the hot BMI resin which behaves like a relatively poor solvent and therefore the PI is dispersed rather than dissolved in the BMI resin, unlike solution blending. As a result a large amount of discrete articles is observed on the neat resin.

7.7 Conclusion

As this chapter has shown, there has been a significant progress in the development of toughened polymers for both thermoplastics and thermoset resins over the last few decades. The increasing understanding of the miscibility requirements, the role of compatibilisers and the toughening mechanisms will further enhance the success in developing blends with tailor-made properties.

The role of modifiers, at the present time, is much clearer in thermoplastics than in thermosets. To successfully exploit these toughened polymers to their full potential, especially in the thermoset resins, further understanding of the relationships between plastics deformations, crosslink density, particle size distributions and the properties of the tougheners are required.

References

1. Paul, D.R. and Barlow, J.W. (1984) A binary interaction model for miscibility of copolymers in blends. *Polymer*, **25**, 487.
2. Manson, J.A. and Sperling, L.H. (1976) *Polymer Blends and Composites*, Plenum Press, New York.
3. Noolandi, J. and Hong, K.M. (1984) Effect of block copolymers at a demixed homopolymer interface. *Macromolecules*, **17**, 1531.
4. Liebler, L. (1982) Theory of phase equilibria in mixtures of copolymers and homopolymers. 2: Interfaces near the consulate point. *Macromolecules* **15**, 1283.
5. Paul, D.R., Locke, C.E. and Vinsa, C.E. (1973) Chlorinated polyethylene modification of blends derived from waste plastics. Part 1: Mechanical behaviour. *Polym. Eng. Sci.* **13**, 202.
6. Locke, C.E. and Paul, D.R. (1973) Graft copolymer modification of polyethylene/polystyrene blends. II: Properties and modified blends. *J. Appl. Polym. Sci.* **17**, 2791.
7. Barentsen, W.M. and Heikens, D. (1973) Mechanical properties of PS/LDPE blends. *Polymer* **14**, 549.
8. Barentsen, W.M., Heikens, D. and Piet, P. (1974) Effect of addition of graft copolymer on the microstructure and impact strength of PS/LDPE blend. *Polymer* **15**, 119.
9. Siqueira, D.F., Galenmbeck, F. and Nunes, S.P. (1991) Adhesion and morphology of PVDF/PMMA and compatibilised PVDF/PS interfaces. *Polymer* **32**, 990.
10. Keskula, and Paul, D.R. (1987) Methyl-methacrylate grafted rubbers as impact modifier for styrenic polymers. *Polym. Sci. Eng. Preprint* **57**, 674.
11. Angola, J.C., Fujita, Y., Sakai, T. and Inoue, T. (1989) Compatibiliser-aided toughening in polymer blends consisting of brittle polymer particles dispersed in a ductile polymer matrix. *J. Polym. Sci., Polym. Phys.* **26**, 807.
12. Choudhary, V., Varma, H.S. and Varma, K. (1991) Polyolefin blends: Effect of EPDM rubber on crystallisation, morphology and mechanical properties of PP/EPDM blend 1. *Polymer* **32**, 2534.
13. Choudhary, V., Varma, H.S. and Varma, K. (1991) Effect of EPDM rubber on melt rheology morphology and mechanical properties of PP/HDPE (90/10) blend 2. *Polymer* **32**, 2541.

14. Van Gisbergen, J.G.M., Borgmans, C.P.J.H., Van der Sanden, M.C.M. and Lemstra, P.J. (1990) Impact behaviour of PS/EPDM-rubber blends: Influence of electron beam irradiation. *Polymer Communications* **31**, 162.
15. Jian, R., Quirk, R.P., White, J.L. and Min, K. (1991) Polycarbonate–polystyrene block copolymers and their application as compatibilising agents in polymer blends. *Polym. Eng. Sci.* **31**, 1545.
16. US Patent 998439.
17. Borggreve, R.J.M., Gaymans, R.J. and Schuijer, J. (1989) Impact behaviour of nylon-rubber blends: 5. Influence of the mechanical properties of the elastomer. *Polymer* **30**, 71.
18. Wroctecki, C., Heim, P. and Gaillard, P. (1991) Rubber toughening of poly(methyl methacrylate). Part 1: Effect of the size and hard layer composition of the rubber particles. *Polym. Eng. Sci.* **31**, 213.
19. Bucknall, C.B. (ed.) (1977) *Toughened Plastics*, Applied Science Publishers, London.
20. Tinker, A.J. (1985) Factors influencing the impact properties of PP/NR blend. *Proc. Int. Conf. Toughening of Plastics—II.* The Plastic and Rubber Institute, 2–4 July, London, Chapter 14.
21. Dao, K.C. (1982) Mechanical properties of polypropylene/crosslink rubber blends. *J. Appl. Polym. Sci.* **27**, 4799.
22. Jang, B.Z., Uhlmann, D.R. and Vander Sande, J.B. (1984) Crystalline morphology of polypropylene and rubber modified polypropylene. *J. Appl. Polym. Sci.* **29**, 4377.
23. Martuscelli, E., Maglio, G., Paalumbo, R., Malinconico, M., Greco, R., Rogosta, G. and Sellitti, C. (1985) Rubber modification of polyamide 6: methods–properties–morphology relationships. *Proc. Int. Conf. Toughening of Plastics-II.* The Plastic and Rubber Institute, 2–4 July, London, Chapter 12.
24. Nadkarni, V.M., Shingankuli, V.L. and Jog, J.P. Blends of thermoplastic polyesters with amorphous polyamide 1. Thermal and crystallisation behaviour. *Polym. Eng. Sci.* **28**, 1326.
25. Teh, J.W. and Rudin, A. (1991) Properties and morphology of polystyrene and linear low density polyethylene polyblend and polyalloy. *Polym. Eng. Sci.* **31**, 1033.
26. Fowler, M.W. and Baker, W.E. (1988) Rubber toughening of PS through reactive blending. *Polym. Eng. Sci.* **28**, 1427.
27. Akhtar, S. and White, J.L. (1991) Characteristics of binary and ternary blends of poly (*p*-phenylene sulfide) with poly(bisphenol A)sulfone and polyetherimide. *Polym. Eng. Sci.* **31**, 84.
28. Weiss, R.A., Huh, W. and Nicolais, L. (1987) Non-reinforced polymer based on blends of polystyrene and thermotropic liquid crystalline polymer. *Polym. Eng. Sci.* **27**, 684.
29. Isayev, A.I. and Modic, M.T.J. (1987) *Polymer Composite* **8**, 158.
30. Nobile, M.R., Amendola, E., Nicolais, L., Acierno, D. and Carfagne, C. (1989) Physical properties of blend of PC and LC copolyesters. *Polym. Eng. Sci.* **29**, 244.
31. Shin, B.Y. and Chung, I.J. (1990) Polymer blend containing a thermotropic polyester with long flexible spacer in the main chain. *Polym. Eng. Sci.* **30**, 22.
32. Isayev, A.I. and Subramanian (1992) Blends of a liquid crystalline polymer with polyether ether ketone. *Polym. Eng. Sci.* **32**, 85.
33. Bucknall, C.B. and Jones, D.P. (1982) Deformation and fracture of thermoplastic containing spherical particles: A comparison between rigid spheres and rubber particles. *Proc. Int. Conf. Deformation, Yield and Fracture of Polymers.* The Plastic and Rubber Institute, Cambridge, Chapter 27.
34. Chong, L.S., Mai, Y.W. and Cotterell, B. (1989) Impact fracture energy of mineral-filled polypropylene. *Polym. Eng. Sci.* **29**, 505.
35. Lee, H. and Neville, N. (1967) *Handbook of Epoxy Resins*, McGraw-Hill Inc.
36. Shimp, D.A. (1986) The translation of dicyanate structure and cyclotrimerization efficiency to polycyanurate properties. *Polym. Mat. Sci. Eng.*, ACS meeting, New York, p. 107.
37. Landman, D. (1986) Advances in the chemistry and applications of BMI in *Developments in Reinforced Plastics—5* Pritchard, G., ed., Elsevier Applied Science Publishers, London and New York, Chapter 2, p. 39.
38. Gillham, J.K., Mcphersin, K.A. and Manzione, L.T. (1981) Rubber modified epoxies: I transition and morphology. *J. Appl. Polym. Sci.* **26** (3), 907.
39. Bucknall, C.B. and Partridge, I.K. (1983) Phase separation in epoxy resins containing PES. *Brit. Polym. J.* **15**, 71.

40. Sultan, J.N., Laible, R.C. and McGarry, F.J. (1971) Microstructures of two phase polymers. *Appl. Polym. Symposium* **16**, 127.
41. Sultan, J.N. and McGarry, F.J. (1969) Microstructural characteristics of toughened thermoset polymers. *MIT School of Engineering Report* **R69-59**.
42. Rowe, A.E., Siebert, A.R. and Drake, R.S. (1970) Toughening thermosets with liquid butadiene/acrylonitrile polymers. *Modern Plastics* **47**, 110.
43. Siebert, A.R. and Riew, C.K. (1971) The chemistry of rubber toughened epoxy resins I. *Proc. 161st ACS meeting, Organic Coatings and Plastics Division*, Los Angeles, CA.
44. Riew, C.K., Rowe, E.A. and Siebert, A.R. (1976) Rubber toughened thermosets in *Toughness and Brittleness of Plastic*, Deanin, R.D. and Crugnola, A.C.S., eds, Adv. Chem. Ser. No. 154, Washington.
45. Bucknall, C.B. and Yoshii, T. (1978) Relationships between structure and mechanical properties in rubber toughened epoxy resins. *Brit. Polym. J.* **10**, 53.
46. Kunz-Douglass, S., Beaumont, P.W.R. and Ashby, M.F. (1980) A model for the toughness of epoxy-rubber particulate composites. *J. Mat. Sci.* **15**, 1109.
47. Bascom, W.D., Cottington, R.L., Jones, R.L. and Peyser, P. (1975) The fracture of epoxy and elastomer-modified epoxy polymers in bulk and as adhesives. *J. Appl. Polym. Sci.* **19**, 2545.
48. Kinloch, A.J., Shaw, S.J., Todd, D.A. and Hunston, D.L. (1983) Deformation and fracture behaviour of a rubber toughened epoxy. 1: Microstructures and fracture studies. *Polymer* **24**, 1341.
49. Yee, A.F. and Pearson, R.A. (1986) Toughening mechanisms in modified epoxies. Part 1: Mechanical studies. *J. Mat. Sci.* **21**, 2462.
50. Kirshenbaum, S.L. and Bell, J.P. (1985) Matrix viscoelasticity:controlling factor in rubber toughening epoxy resins. *J. Appl. Polym. Sci.* **30**, 1985.
51. Lee, W.H., Hodd, K.A. and Wright, W.W. (1985) The influence of crosslink density on the toughness of rubber modified epoxy resins. *Proc. Int. Conf. Adhesives, Sealants and Encapsulant—1*, Network Events Ltd, Plastic and Rubber Institute, London, p. 144.
52. Pocius, A.V. (1987) Third generation 2-part epoxy adhesives. *19th International SAMPE Tech. Conf.*, Oct 13–15, p. 312.
53. Lee, W.H. (1986) *Elastomer modified epoxy resins (toughening of TGDDM-DDS epoxy)* Ph.D Thesis, Brunel University of West London.
54. British Patent Application No. 8928181.
55. Yang, P.C., Woo, E.P., Bishop, M.T. and Pickelmaan, D.M. (1990) Rubber toughening of thermosets—A system approach. *Polym. Mat. Sci. Eng.* (preprints) **63**, 315.
56. Shaw, S.J. and Kinloch, A.J. (1984) High temperature adhesives. *Proc. Int. Adhesion Conf.*, The Plastics and Rubber Institute, Chapter 3.
57. Stanzenberger, H., Konig, P., Herzog, M. and Romer, W. (1987) Toughened bismaleimides: Concepts, achievements, directions. *Proc. 19th Int. SAMPE Tech. Conf.*, October, Virginia, p. 372.
58. Hunston, D.L., Moulton, R.J., Johnston, N.J. and Bascom, W.D. (1987) Matrix resin effects in composite delamination: Mode 1 fracture aspects in *Toughened Composites* Johnston, N.J., ed., Symposium on Toughened Composite, ASTM. STP937, p. 74.
59. Lee, W.H. (1989) The influence of neat resin properties in composites. *Proc. 1st Japan Int. SAMPE symposium*, Tokyo, Nov. 1989, p. 576.
60. Raghava, R.S. (1988) Development and characterisation of thermosetting-thermoplastic polymer blends for application in damage tolerence composites. *J. Polym. Sci., Polym. Phys.* **26**, 65.
61. Raghava, R.S. (1987) Role of matrix-particle interface adhesion on fracture toughness of dual phase epoxy-polyethersulfone blends. *J. Polym. Sci., Polym. Phys.* **25**, 1017.
62. Bucknall, C.B. and Gilbert, A.H. (1989) Toughening of tetrafunctional epoxy resins using polyetherimide. *Polymer* **30**, 213.
63. Hourston, D.J. and Lane, J.M. (1992) The toughening of epoxy resins with thermoplastics: 1. Trifunctional epoxy resin–polyetherimide blends. *Polymer* **33**, 1379.
64. Hedrick, J.L., Yilgor, I., Hedrick, J.C., Wilkes, G.L. and McGrath, J.E. (1985) *Proc. 30th National SAMPE Symposium*, Covina, California, March 19–21, p. 947.
65. Yamanaka, K. and Inoue, T. (1989) Structure development in epoxy resin modified with poly-(ether sulphone). *Polymer* **30**, 662.

66. Cecera, J.A. and McGrath, J.E. (1986) Morphology and properties of amine terminated poly(arylether ketone) and poly(arylether sulphone) modified epoxy resin systems. *Polymer* (preprints), **27**, 299.
67. Shimp, D.A., Christenson, J.R. and Ising, S.J. (1990) *Cyanate Ester Resins—Chemistry, Properties and Applications.* Hi-Tek Polymers, Inc., Louisville.
68. Cantwell, W.J. and Roulin-Moloney, A.C. (1989) Fractography and failure mechanism of unifilled and particulate filled epoxy resins in *Fractography and Failure Mechanisms of Polymers* Roulin-Moloney, A.C., ed., Elsevier Applied Science Publishers, Chapter 7, p. 244.
69. Spanoudakis, J. and Young, R.J. (1984) Crack propagation in a glass particle filled epoxy resin. Part 2: Effect of particle matrix adhesion. *J. Mat. Sci.* **19**, 487.
70. Kinloch, A.J., Maxwell, D.L. and Young, R.J. (1985) The fracture of hybrid-particulate composites. *J. Mat. Sci.* **20**, 4169.
71. Low, I.M., Mai, Y.W., Bandyopadhay, S. and Silva, V.M. (1987) in *Proc. 1987 Australian Fracture Group Symposium.* Mai, Y.W., ed., Sydney, Australia, p. 77.
72. Galbriath, S.T., McClusky, J., Orton, M.L. and Nield, E. (1985) Morphology and properties of interpenetrating network derived from polyurethane and polyethacrylates. *Proc. Int. Conf. Toughening of Plastics—II.* The plastics and Rubber Institute, July, London.
73. Hartness, J.T. (1985) A dicyanate semi-IPN (SIPn) matrix composite in *Toughened Composites,* Johnston, N.J., ed., STP 937, p. 453.
74. Sefton, M.S., McGrail, P.T., Peacock, J.A., Wilkinson, S.P., Crick, R.A., Davies, M. and Almen, G. (1987) Semi-interpenetrating polymer networks as a route to toughening of epoxy resin matrix composites, Lynch, T., Presh, J., Wolf, T. and Rupert, N., eds., *Proc. 19th Int. SAMPE Tech. Conf.* October 13–15, Virginia, p. 700.
75. Hanky, A.O. and St Clair, T.L. (1986) *SAMPE J.* **21**, 40.
76. Pater, R.H. (1991) Interpenetrating polymer network approach to tough and microcracking resistance high temperature polymers. Part II, Larc-RP41. *Polym. Eng. Sci.* **31** (1), 20.
77. Lee, W.H., Hodd, K.A. and Wright, W.W. (1992) Plastic deformation in rubber modified crosslink epoxy. *J. Mat. Sci.* **27**, 4582.
78. Sternstein, S.S. and Ongchin, L. (1969) Yield criteria for plastic deformation of glassy high polymer in general stress fields. *ACS Polymer Preprints* **10** (2) 1117.
79. Kambour, R.P. (1964) Structures and properties of crazes in polycarbonate and other glassy polymers. *Polymer* **5**, 143.
80. Bucknall, C.B. and Smith, R.R. (1965) Stress-whitening in high impact polystyrene. *Polymer* **6**, 437.
81. Bucknall, C.B., Clayton, D. and Keast, W.E. (1972) Rubber toughened plastics—Part 2: Creep mechanisms in HIPS/PPO. *J. Mat. Sci.* **7**, 1443.
82. Donald, A.M. and Kramer, E.J. (1982) The competition between shear deformation and crazing in glassy polymers. *J. Mat. Sci.* **17**, 1871.
83. Kambour, R.P. (1973) A review of crazing and fracture in thermoplastics. *J. Polym. Sci., Macromolecular Reviews* **7**, 1.
84. Donald, A.M. and Kramer, E.J. (1982) Interaction of crazes with pre-existing shear bands in glassy polymers. *J. Mat. Sci.* **17**, 1739.
85. Melander, A. and Staklberg, U. (1980) The effects of void size and distribution on ductile fracture. *Int. J. Fracture* **16**, 431.
86. Needleman, A (1972) Void growth in an elastic-plastic medium. *J. Appl. Mech., Trans, ASME,* 964.
87. Gurson, A.L. (1977) Continum theory of ductile rupture by void nucleation and growth. Part 1: Yield criteria and flow rules for porous ductile media, *J. Eng. Mat. Tech., Trans., ASME,* **99**, 2.
88. Yamamoto, H. (1978) Conditions for shear localisation in the ductile fracture of void containing materials. *Int. J. Fracture* **14**, 347.
89. Trevgaard, V. (1981) Influence of voids on shear band instabilities under plain strain conditions. *Int. J. Fracture* **17**, 389.
90. Hobbs, S.Y. (1977) Fracture toughness of PC structural foams. *J. Appl. Phys.* **48**, 4052.
91. Yee, A.F. (1977) The yield and deformation behaviour of some polycarbonate blends. *J. Mat. Sci.* **12**, 757.
92. Yee, A.F. and Kambour, R.P. (1978) Fracture toughness studies on rubber toughened polymers. *Proc. Int. Conf. Toughening of Plastics,* The Plastic and Rubber Institute, London.

93. Maxwell, M.A. and Yee, A.F. (1981) The effects of strain rate on the toughening mechanisms of rubber modified plastics. *Polym. Eng. Sci.* **21**, 205.
94. Boggreve, R.J.M., Gaymans, R.J. and Eichenwald, H.M. (1989) Impact behaviour of nylon-rubber blends. 6: Influence of structure on voiding processes, toughening mechanism. *Polymer* **30**, 78.
95. Gent, A.N. and Lindley, P.B. (1958) Internal rupture of bonded rubber cylinders in tension. *Proc. Roy. Soc.* **A249**, 195.
96. Huang, Y. and Kinloch, A.J. (1992) The toughness of epoxy polymers containing microvoids. *Polymer* **33** (6), 1330.
97. Evans, A.G., Ahmad, Z.B., Gilbert, D.G. and Beaumont, P.W.R. (1985) Mechanisms of toughening in rubber toughened polymers. *Proc. Int. Conf. Toughening of Plastics—II*. The Plastic and Rubber Institute, 2–4 July, London, Chapter 16.
98. Lange, F.F. (1970) The interaction of a crack front with a second-phase dispersion. *Phil. Mag.* **22**, 983.
99. Evans, A.G. (1972) The strength of brittle materials containing second phase dispersions. *Phil. Mag.* **26**, 1327.
100. Green, D.J. (1979) Fracture of a brittle particulate composite. Part 2: Theorectical aspets. *J. Mat. Sci.* **14**, 1657.
101. Donald, A.M. and Kramer, E. (1982) Plastic deformation mechanisms in polyacylonitrile-butadiene styrene (ABS). *J. Mat. Sci.* **17**, 1765.
102. Dillon, M. and Bevis, M.J. (1982) The microstructure and deformation of model ABS compounds. Part 2: The effect of graft frequency and rubber particle size. *J. Mat. Sci.* **17**, 1903.
103. Sultan, J.N. and McGarry, F.J. (1973) Effect of rubber particle size on deformation mechanism in glassy epoxy. *Polym. Sci. Eng.* **13**, 29.
104. Bascom, W.D., Ting, R.Y., Moulton, R.J., Riew, C.K. and Siebert, A.R. The fracture of an epoxy polymer containing elastomeric modifiers. *J. Mat. Sci.* **16**, 2657.
105. Bevis, M.J. and Allen, P.S. (1987) Multi-live-feed injection moulding. *Plast. Rubb. Proc. Appln.* **7**, 3.
106. Ibrahim, A.M., Quinlivan, T.J. and Saferis, J.C. (1986) Processing of PES reinforced high performance epoxy blends. *Polymer Preprints* **27**, 277.

8 Blends containing liquid crystal polymers

C.S. BROWN and P.T. ALDER

8.1 Introduction

Liquid crystal polymers (LCPs) are used in many different commercial applications ranging from high modulus fibres (Kevlar) to microwave cookware (Xydar). Much of the driving force behind the development of LCPs had been the increasing desire to replace metal components with engineering polymers, for example in automotive and aerospace sectors. LCPs were proclaimed as a new class of engineering polymers, which would take a large share of this market. In the event they have achieved limited market penetration as engineering materials in their own right. Key properties turned out to be low thermal expansion coefficients leading to high quality mouldings, rather than exceptional modulus or strength. Suppliers have been withdrawing production from the area (e.g. ICI, Eastman Kodak). Over the same period commercial interest has grown in polymer blend technology. Thus it is not surprising that the focus of interest in LCPs has switched from their use as stand alone materials to their use in blends.

This chapter focuses on recent developments in the area of blends containing LCPs. A brief introduction to liquid crystals and LCPs is provided before the potential applications of LCP blends are discussed.

8.1.1 The nature of liquid crystals

A liquid crystal is quite simply a state of matter between the liquid and the crystal state. A liquid crystal is an *ordered fluid*. The possibility of order existing, at equilibrium, in a fluid arises when the molecules within the liquid are anisotropic. These anisotropic molecules must also resist entropic forces which would otherwise tend to turn a rod into a random coil. Rigid rods readily form liquid crystal phases. Rigidity is, however, not as strong a requirement as originally perceived. Several so-called 'rigid rod' molecules have in fact been shown to be close to random coils in solution [1]. Other factors, such as the preferred conformation of the chain, can prevent an anisotropic molecule collapsing into a random coil [2].

Almost any form of anisotropy can give rise to a liquid crystal phase (see Figure 8.1). Rods, discs, planks and helices are all well documented examples [3].

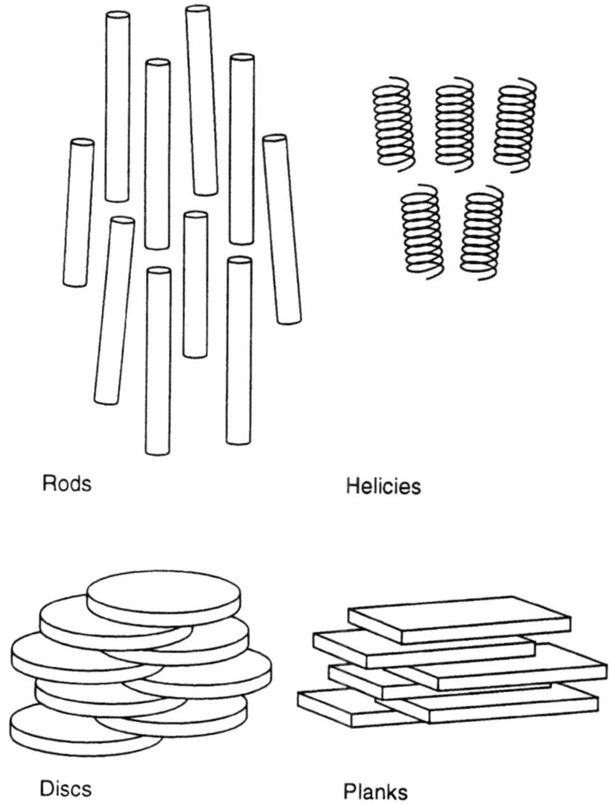

Figure 8.1 Anisotropic units giving rise to liquid crystal phases.

Many natural materials can exist in the liquid crystal state. Examples of common materials include DNA, cholesterol and tobacco mosaic virus. Probably the best known synthetic liquid crystals are those used in liquid crystal displays. Gray *et al.* [4] pioneered the breakthrough in this field with the development of cyano-biphenyls. Other common structural units include; para-linked benzene rings, cyclohexane, bi-cyclo [2, 2, 2] octane, stilbene, azo-links, etc. A comprehensive reference to liquid crystals is Kelker [5] which lists many of the chemical classes of material that exhibit liquid crystallinity.

There are many different types of liquid crystals but there are three main classifications as recognised by Friedel [6]: nematic, smectic and cholesteric (see Figure 8.2). The nematic state is ordered in one dimension only and is the least ordered liquid crystal phase. The smectic phase contains some form of two-dimensional order, such as layers, hence the term 'smectic'. Cholesteric liquid crystals are twisted structures usually due to an optically active molecule. Cholesterol itself was the first example of these twisted structures, hence the name (see Marchessault *et al.* [7]).

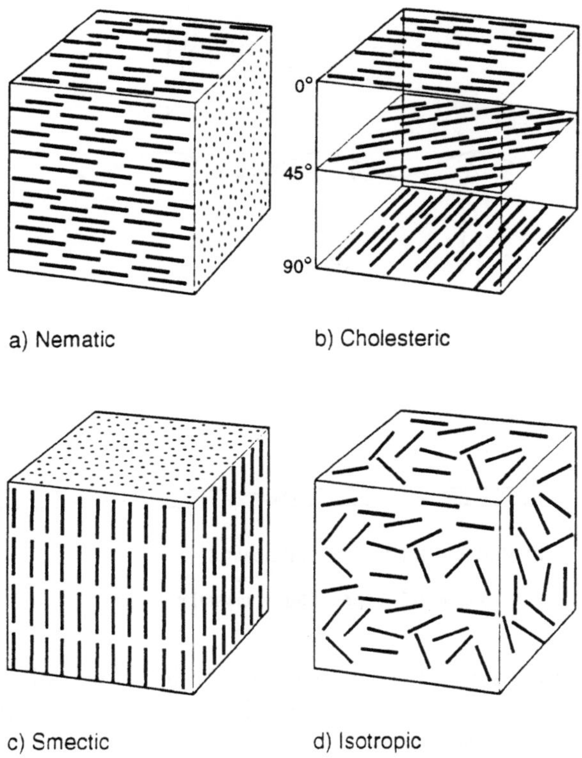

Figure 8.2 Main classifications of liquid crystal phases.

A crystal can become a random liquid in two ways: by melting or dissolving. In the same way the phase change from crystal to liquid crystal can either be achieved by heating in which case it is called 'thermotropic', or by dissolution, in which case it is called 'lyotropic'. The basic anisotropic unit in lyotropic systems is often not a single molecule but an agglomeration. Soaps, for example, can form lyotropic liquid crystal phases [8].

8.1.2 Liquid crystal polymers

A LCP is simply a polymer that exhibits one of the above liquid crystal structures. There are two main ways that anisotropic building blocks can form a polymer. They can either be joined end to end to form a *main chain LCP* or they can dangle off the side of a normal polymer chain to form a *side chain LCP* (Figure 8.3). Anisotropic aromatic building blocks can very usefully be copolymerised with traditional flexible monomers. If these flexible units are placed within the main chain they can be used to control the melting point and glass transition temperature of the LCP. Flexible units can also be placed

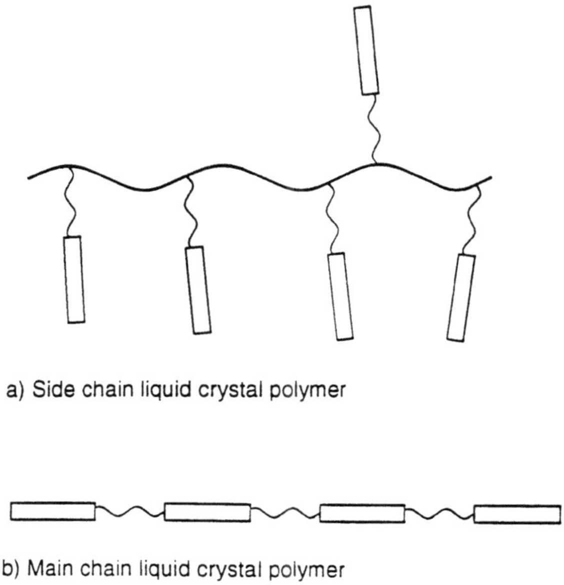

Figure 8.3 Types of liquid crystal polymers.

between a normal polymer backbone and the liquid crystal side chain to decouple the motion of the liquid crystal group from the main chain. An interesting conflict arises here because the flexible chain will tend to want to become a random coil and the liquid crystal groups want to become ordered, as illustrated in Figure 8.4.

Many other combinations of liquid crystal structures with polymer architectures are being explored. For example, it is possible to incorporate liquid crystal groups into an elastomer [3]. Further intriguing opportunities then arise such as colour changes on stretching.

The most important LCPs for blending with bulk polymers are thermotropic main chain LCPs that exhibit a liquid crystal phase within the processing window of the bulk polymer but which then solidify on cooling to room temperature. As with any polymer a LCP can either crystallise on cooling or go through a glass transition temperature. Hence, at room temperature, the structure of the LCP is that of an *ordered* glass or a semi-crystalline polymer. The degree and type of order can vary enormously depending on chemical structure or thermal history. The exact description of the order and how it arises, particularly in copolymers, is still a subject of academic debate.

The chemistry of commercial LCPs has been dominated by aromatic polyamides (e.g. Du Pont's Kevlar) and aromatic polyesters (e.g. Hoechst-Celanese's Vectra or Amoco's Xydar). However, many other chemical types have been synthesised (e.g. liquid crystal polycarbonates, liquid crystal polyphosphazines and liquid crystal siloxanes [3]). The use of commercially

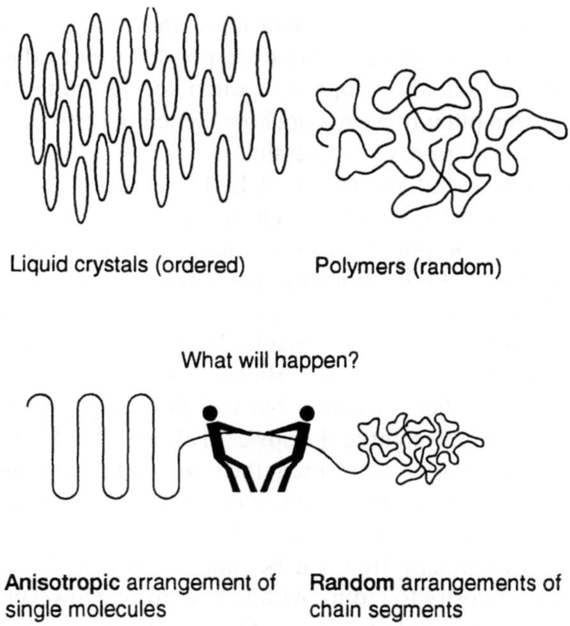

Figure 8.4 The liquid crystal polymer conflict of interests between the tendency towards an ordered fluid (arising from the liquid crystals) and the tendency towards a random coil (arising from the polymer) (after Engel *et al.* [9]).

available liquid crystal materials in blends with bulk polymers has been investigated quite extensively (but by no means exhaustively) over the last few years. However, before discussing the benefits that may arise from blending LCPs with other polymers it is useful to discuss what special properties LCPs possess and what materials are available.

8.1.3 Main properties of LCPs

As LCPs are often formed from highly aromatic stiff monomers it is hardly surprising that they can form very stiff polymers. Kevlar fibres, for example, have a specific modulus considerably in excess of steel [10,11]. Carbon fibres with moduli up to 700 GPa can be produced from a liquid crystal pitch [12]. The advantage that the liquid crystal phase brings is that processing from a naturally ordered fluid gives rise to a more perfect final structure than processing a normal chaotic fluid. These highly aromatic LCPs are also very brittle, have very low failure strains, but can be very strong.

The mechanical properties of LCPs are highly anisotropic. This does not present a significant problem in LCP fibre applications but is a serious problem for most other applications. Injection moulded specimens can show anisotropies in modulus of up to 20, for example [13]. There is also a strong skin core effect which makes the properties at the surface very different to the bulk. The industry in general has not responded well to the challenge of harnessing the inherent anisotropy that often results from LCPs. Instead much effort has been put into producing homogeneous, isotropic components, which in some cases has proved successful.

One of the most useful properties of LCP mouldings is their very low thermal expansion coefficients. This results in low mould shrinkage and hence very high tolerance components can be moulded. This is one of the most exploited properties of LCPs. For example, intricate electronic components that could not be moulded from traditional polymers can be made from LCPs. This benefit is significant enough to justify the extra material cost.

Another property of LCPs that is particularly important when considering their use in blends is their viscosity. Normal flexible polymers increase in viscosity with molecular weight to the power of about 3.4 [14]. If the LCPs are aligned in the direction of flow then their viscosity only increases with molecular weight to the power 1. Thus, if equal molecular weights are compared, LCPs have much lower viscosities. In fact the viscosity of a LCP can even increase with temperature (as it becomes a normal fluid) whereas the usual trend is for viscosity to reduce with temperature. However, some of the more ordered types of liquid crystals, i.e. the smectics, tend to have very high viscosity as they are starting to possess many of the properties of solids.

The understanding of the viscosity of both liquid crystals and polymers has increased significantly over the last decade or so. The liquid crystal display market gave great impetus to the study of the viscosity of liquid crystals [15]. Doi and Edwards [16] realised the significance of entanglements to polymer flow and succeeded in modeling viscosity in terms of a tube model. However, the current level of understanding of the combined field of the flow of LCPs is primitive by comparison. The study of the flow of blends of LCPs with other polymers is only just beginning.

8.1.4 Commercially available LCPs

Commercially available main chain LCPs are listed in Table 8.1. Table 8.2 gives some typical properties. Table 8.3 highlights some LCP blend systems that have been studied.

8.2 The effect of LCPs on the processing of polymers

One of the most attractive features of LCPs is their ability to alter the rheology of bulk polymers. The way in which LCPs alter bulk polymer flow is not yet well understood. There is a wide range of effects reported in the academic literature on an increasing number of systems. Most reports in the literature are concerned with viscosity reduction.

Table 8.1 Examples of liquid crystal polymers and their physical properties.

LCP name	Manufacturer	Formulation	M.pt. (°C)	T_g(°C)	T_{proc}(°C)
Vectra A	Hoechst Celanese	HNA (27%), HBA (73%)	283	105	320–340
Vectra B	Hoechst Celanese	HNA (58%), TPA (21%), 4AP (21%)	280	110	300–330
X7-range*	Eastman Kodak	PET/HBA (HBA 40, 60 or 80%)	250–300	70,143	270–350
Rodrun	Unitika	PET/HBA based materials	250?	70,143	270–350
LX-series	Du Pont	TA, HQ, PHQ	310	180	340–360
Ultrax KR*	BASF	?	—	195?	270–320
Xydar†	AMOCO	BP, HBA, TA	412	—	>420
Victrex SRP*	ICI	HBA, HQ, IA	280	—	300–350
KU-9211*	Bayer	HBA, TA, HQ, IA, BP	—	198	330–360

*Currently withdrawn
†Second generation LCP formulations available with processing temperature < 350 °C.

HNA	2-hydroxy-6-naphthoic acid	HBA	4-hydroxybenzoic acid
4AP	4-aminophenol	PET	poly(ethylene terephthalate)
TA	terephthalic acid	HQ	hydroquinone
PHQ	phenyl hydroquinone	BP	4,4′-dihydroxybiphenyl
IA	isophthalic acid		

Table 8.2 Selected mechanical properties of LCPs.

Property	Vectra A950	Vectra B950	Ultrax KR-4002	KU 1-9211	LX-2000
Tensile strength (MPa)	165	188	160	160	117
Tensile modulus (GPa)	9.71	9.3	8.2	9.95–16.0	28.3
Tensile elongation (%)	3.0	1.3	2.8	1.7–2.8	0.6
Flexural modulus (GPa)	9	15.2	—	6.2–10.0	18
Notched izod impact strength (J/m)	520	415	—	—	197
(kJ/m^2)	—	—	59	32	—
Heat deflection temperature (°C)	180	—	118	156	185

Table 8.3 Examples of LCP blends that have been investigated (after Zaldua [17])

Components of the blends[a]	Studied properties
P(HBA–ETP)/PBT	Compatibility by DSC
P(HBA–ETP)/PC	Morphology, rheology
P(HBA–ETP)/PC	Morphology, nmr, transreactions
P(HBA–ETP)/PC	Rheology, DSC, dielectric, mechanical
P(HBA–ETP)/PC	Rheology, shrinkage, morphology
P(HBA–ETP)/PET	DSC, morphology, mechanical
P(HBA–ETP)/PET	Drawing, dynamic viscoelasticity, DSC
P(HBA–ETP)/PET	DSC, mechanical properties
P(HBA–ETP)/PET	Rheology, DSC, dielectric, mechanical
P(HBA–ETP)/PET	Phase separation
P(HBA–ETP)/nylon	Morphology, rheology
PHBA–PET/CPVC	Rheology, spiral mould flow
PHBA–PET/PS	Rheology, DSC, dielectric, mechanical
P(HBA–ETP)/PHMT	Interaction between components
P(HNA–HBA)/PA	Rheology, morphology, mechanical
P(HNA–HBA)/PC	DSC, spinning, shrinkage, dynamic viscoelasticity
P(HNA–HBA)/PC	Rheology, mechanical
P(HNA–HBA)/PC	Morphology
P(HNA–HBA)/PC	Morphology, mechanical, rheology
P(HNA–HBA)/PET	Morphology, mechanical
Naphthalene based LCC/PES: PAr: PEEK	Morphology, processing, mechanical
PC: PBT: nylon Hoechst Celanese	Processability, mechanical
Polyesteramide/nylon TBBA/PS	DSC, shrinkage
EBBA/PVC	Permeability, DSC, morphology
PB8/PBT	DSC
Polyesteramide/PBT	DSC

[a] P(HBA–ETP): 60% *p*-hydroxybenzoic acid–40% poly(ethylene terephthalate); PBT: poly(butylene terephthalate); PC: polycarbonate; CPVC: chlorinated poly(vinyl chloride); PHMT: poly(hexamethylene terephthalate); PA: polyamide; P(HNA–HBA): poly(2-hydroxy-6-naphthoic acid)–poly(*p*-hydroxybenzoic acid); LCC: liquid crystalline copolyster; PES: poly(ether sulphone); PAr: polyarylate; PEEK: poly(ether ether ketone); TBBA: terephthal bis-4-*n*-butylaniline; EBBA: *N*-(4-ethoxybenzylidene)-4′-butylaniline; PB8: poly(bisphenyl-4,4′-ylene sebacate).

8.2.1 Examples of rheological changes

Blizard and Baird [18] reported a viscosity reduction of over an order of magnitude. They blended 30% of a liquid crystal copolyester containing hydroxybenzoic acid and ethylene terephthalate with 70% polycarbonate at 260 °C. The viscosity, as measured by capillary rheometry, reduced from 1000 Pa s for pure polycarbonate to about 20 Pa s at 30% LCP loading (Figure 8.5). Blends of higher LCP concentration were of even lower viscosity,

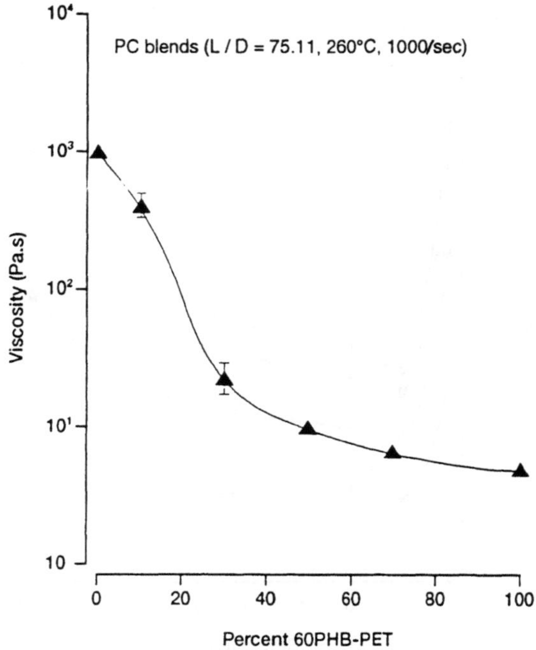

Figure 8.5 Viscosity reduction of polycarbonate (PC) with the addition of a liquid crystal copolyester (60 PHB-PET) (after Blizard and Baird [18]).

but the rate of viscosity reduction was less steep. Siegmann *et al.* [19] observed a steep viscosity reduction when a hydroxybenzoic acid hydroxynaphthoic acid liquid crystal copolyester was blended with an amorphous polyamide (Figure 8.6). In this case the greatest reduction was seen at 5% LCP concentration. With the addition of more LCP the viscosity in fact increased. In some cases the viscosity of the pure LCP was almost the same as the pure polyamide, yet with appropriate mixing the blend can achieve a viscosity nearly an order of magnitude lower. Several further examples of viscosity reduction can be found in the review by Dutta *et al.* [20].

Viscosity increases can also occur in some circumstances. For example, a 20% concentration of a hydroxybenzoic acid/hydroxynaphthoic acid liquid crystal copolymer in nylon 12 increased viscosity at 270–290 °C. Weiss *et al.* [21] also reported a rise in viscosity in a blend of the liquid crystal 4,4′-dihydroxydimethylbenzalazine in polystyrene, but only at low shear rates ($1\,s^{-1}$).

Kulichikhin *et al.* [22] noticed three different types of behaviour in one LCP blend. They measured the viscosity of polysulphone (Amoco's P-3500) blended with a liquid crystal polyester (BASF's Ultrax-4002). At 280 °C log η appeared to fall linearly with LCP content throughout the composition range. At 260 °C the fall was initially faster but increased at higher LCP concentra-

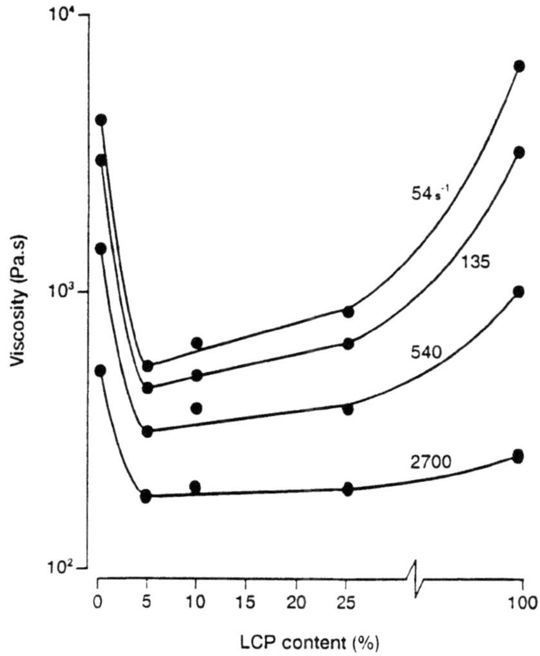

Figure 8.6 Composition dependence of the viscosity of an amorphous polyamide with % LCP (hydroxybenzoic acid/hydroxynaphthoic acid copolymer) at 260 °C at 4 shear rates (after Siegmann *et al.* [19]).

tion giving a minimum value at about 50%. At 240 °C two minima are reported (Figure 8.7).

Nobile *et al.* [23] observed that viscosity can increase or decrease in the same LCP blend at the same temperature (Figure 8.8). They investigated a blend of a bisphenol-A polycarbonate and a liquid crystal copolyester containing ethylene terephthalate and oxybenzoate units. At very low shear rate ($10^{-2}\,s^{-1}$) the viscosity was found to increase with LCP content whereas at high shear rate ($700\,s^{-1}$) a significant drop was observed (about a factor of 10 at 50% LCP content).

From the above examples it is clear that LCPs offer the blend technologist considerable possibilities for the control of viscosity, and hence improving processing, if the behaviour of the blend system can be understood and relied on to be consistent in real polymer processing equipment.

8.2.2 Discussion of viscosity modifications

One consistent theme in the study of LCP blend melts is that fully homogeneous single phases are not observed. Two separate phases are the norm. The morphology of the melt is believed to be very important in determining its

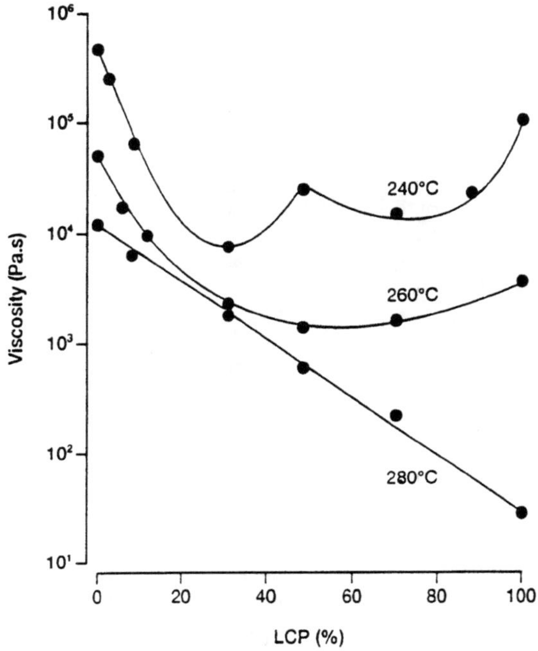

Figure 8.7 Viscosity versus composition for polysulphone blended with a liquid crystalline polyester (ultrax) (after Kulichikhin *et al.* [22]).

rheology. The morphology present in the melt (i.e. droplets, ellipsoids, fibres or layers) is obviously determined by the previous processing that has been applied to the blend. Processing history can also be expected to alter the orientation and domain structure of the liquid crystal component. Certain morphologies may also be transient. For example, La Mantia *et al.* [24] observed that fibrils of Vectra in nylon 6 can relax to droplets during the flow through a long capillary.

The viscosity actually measured (and indeed its strain rate dependence) will thus be a strong function of the previous stress applied to the melt and will also depend on *when* the measurement is made relative to previous processing or thermal history.

If there is no (or very little) prior processing, the LCP will tend to form spherical droplets within the matrix polymer. On average within a droplet there will be no preferred orientation of the LCP molecules, however on a finer scale the LCPs will form into domains which are oriented. The viscosity at this point should fit a model of spheres of one liquid suspended in another (e.g. Einstein-type fluid). During flow these spheres can be expected to deform.

The case of deformable droplets of one fluid in another has been calculated

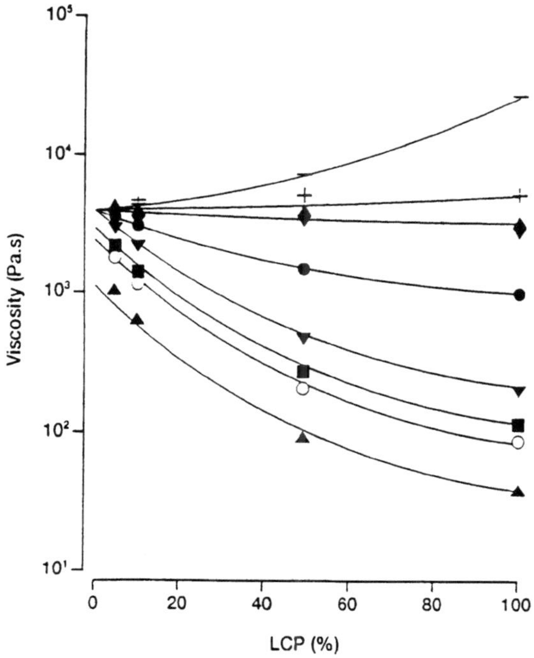

Figure 8.8 Viscosity versus LCP percentage at 240 °C at different shear rates: —, $\gamma = 10^{-2}\,s^{-1}$; +, $\gamma = 1.7 \times 10^{-1}\,s^{-1}$; ♦, $\gamma = 3.1 \times 10^{-1}\,s^{-1}$; ●, $\gamma = 1\,s^{-1}$; ▼, $\gamma = 12\,s^{-1}$; ■, $\gamma = 57\,s^{-1}$; ○, $\gamma = 130\,s^{-1}$; ▲, $\gamma = 700\,s^{-1}$. The LCP is a copolyester containing ethylene terephthalate and oxybenzoate and the bulk polymer is a polycarbonate (after Nobile *et al.* [23]).

by Schowalter *et al.* [25] using the extended Einstein equation:

$$\eta = \eta_1[1 + (5\eta_1/\eta_2 + 2)/(2\eta_1/\eta_2 + 1)\cdot\phi_2]$$

where η_1 is the viscosity of component 1, η_2 is the viscosity of component 2, and ϕ_1 and ϕ_2 are the respective volume fractions.

This equation does not allow for a reduction in blend viscosity. For all compositions and viscosity ratios the viscosity of the blend should increase. It is hard to see how the physics of the situation could change dramatically if this model is extended to ellipsoids or if fibres are considered instead of deformed spheres. Thus this model is not particularly helpful.

When a blend melt containing droplets begins to flow several things can happen. The droplet can deform into a fibre (see section 8.3), the fibre could break up into droplets again as described by Utracki [26] or the droplets could coalesce. It is even possible that fibres could coalesce to form platelets.

Krasnikova *et al.* [27] related the viscosity ratio and applied shear stress to the tendency to form droplets, fibres or melt fracture for blends of two isotropic fluids (polyethylene and polystyrene). This is illustrated in Figure 8.9. The extreme case of perfect cylindrical layers flowing through a circular die has

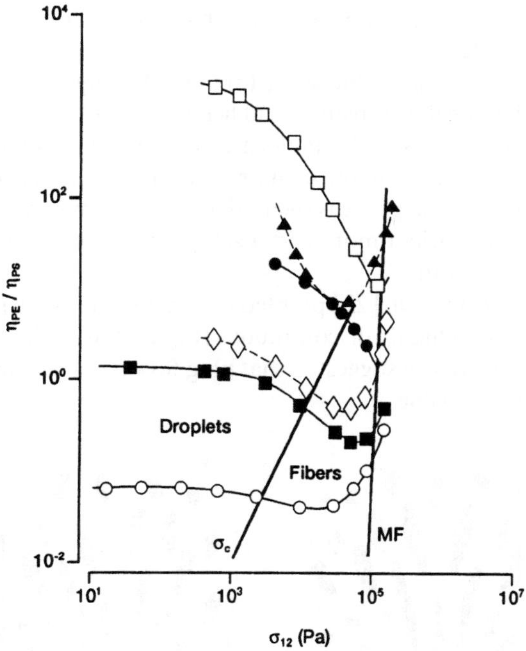

Figure 8.9 Ratio of shear viscosity $\lambda = \eta PE/\eta PS$ for $\phi_{PE} \leqslant 0.3$ versus shear stress. Points: experimental = solid straight line; σ_c = boundary between drops and fibre dispersion regions, MF = melt fracture (PE = polyethylene, PS = polystyrene, σ_{12} = shear stress, ϕ = volume fraction) (after Krasnikova *et al.* [27]).

been calculated by Lees [28]:

$$1/\eta = \phi_1/\eta_1 + \phi_2/\eta_2$$

This equation can in fact explain large viscosity reductions with small LCP addition levels if $\eta_1 \gg \eta_2$. The possibility of forming such layers at high LCP concentration is mentioned by Lee [29, 30]. Delleuse and Jaffa [31] observed such a behaviour in a blend containing two LCPs of different composition. They also found that if the copolymer composition was close enough the LCP/LCP blend was miscible and the usual miscible blend additively law applied:

$$\ln \eta = \phi_1 \ln \eta_1 + \phi_2 \ln \eta_2$$

The phenomena observed in Figures 8.5–8.8 where the viscosity of the bulk polymer is reduced by the lower viscosity LCP, can at least qualitatively be explained by invoking polymer blend theory by treating the two fluids as isotropic fluids approaching Lees layers. However, the general phenomena of a blend viscosity ≪ viscosity of both its components cannot be explained by the traditional theories discussed above. Three observations may be relevant in explaining this effect.

First, LCPs show considerable shear thinning, the liquid crystal molecules become oriented in the flow direction and hence show a lower viscosity. When the applied shear is released the molecules will remain oriented in the low viscosity state. This shear thinning memory effect may well be accentuated in a blend. As the liquid crystal phase is sheared its viscosity may reduce more than the bulk, it may then orient more and a self-perpetuating viscosity reducing mechanism may be set up.

Second, Kulichikhin *et al.* [22] pointed out that the velocity profiles within a LCP blend might be highly discontinuous (Figure 8.10). They felt, however, that the weight of evidence suggested that plug flow (i.e. a continuous velocity profile) was more probable.

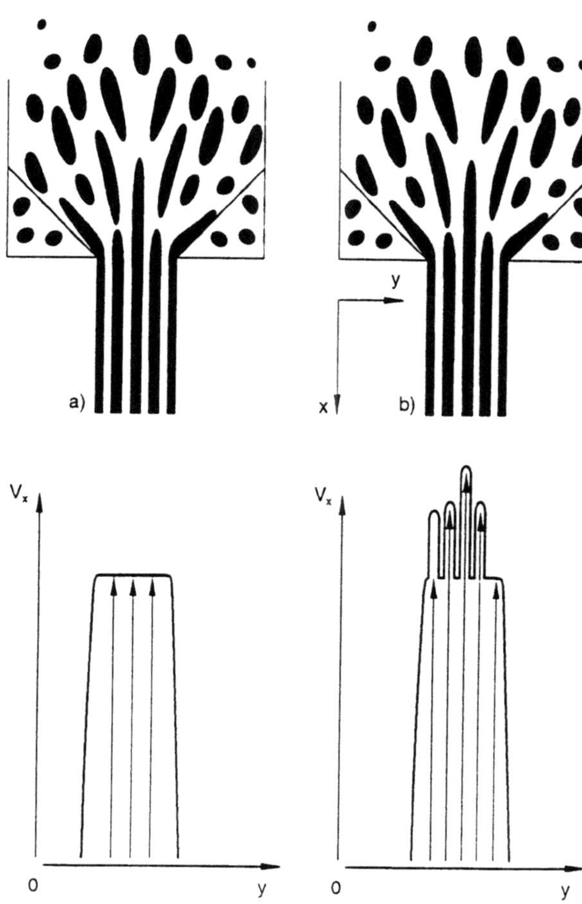

Figure 8.10 A scheme explaining the appearance of the conical fibrillation envelope at the channel entrance and variants of velocity profiles (see text) (after Kulichikhin *et al.* [22]).

Third, the vast majority of work on the rheology of LCP blends reports only shear viscosity. Berry *et al.* [32] pointed out that the elongational viscosity of LCPs and their blends can be 100 to 400 times greater than their shear viscosity. Most practical polymer processing involves a considerable elongational component to the flow, e.g. at the entry of an extruder die. They concluded that elongational flow rather than the shear component dominates the die entry pressure drop. Thus the measured shear viscosities may in fact be the result of differences in elongational viscosity. The importance of elongational flow producing a desirable morphology was noted by Dutta *et al.* [20]. They considered that the extensional deformation of the LCP domains during flow in the entrance region of the viscometer explained the lowering of the viscosity at high shear rates.

8.2.3 Implications of rheology modification

As can be seen from the above discussion, the rheology of LCP blends is a complex but exciting new field of academic interest where many significant questions remain unanswered. There are, however, enough reliable academic reports of significant rheological effects for the industrial processor to take a serious interest in the commercial potential of LCP blends and a number of investigations of LCP blend rheology in semi-commercial processing equipment have been made. For example, Lee [29, 30] has injection moulded a blend of chlorinated PVC and hydroxybenzoic acid/ethylene terephthalate liquid crystal copolymer. He found that the LCP increased the spiral mould flow length and was able to model the improvement in injection moulding processing using a simple power law. He also noted that extrudate die swell was reduced by the addition of the LCP. Toshikazu [33] investigated LCP/ethylene vinyl-alcohol copolymer blends in a 75 ton injection moulding machine and a 20 cm diameter extruder. He observed decreased screw torque and melt pressure in all blend systems and established a relationship between morphology, viscosity and surface tension. This work aimed to establish optimum gas barrier performance from injection moulded parts.

Dutta *et al.* [20] have also commented on the implications of rheology modifications arising from LCPs. They highlighted the benefit of reduced melt temperature which may imply reduced energy costs and/or less degradation. They also pointed out the benefit of reduced viscosity with respect to filling large or complex moulds.

The above discussion has concentrated on viscosity modification as an example of a processing improvement that can be obtained from LCPs. Other processing advantages arising from LCPs have also been reported in the literature: processing temperature can be reduced [20, 34] die swell can be less pronounced with a suitable LCP [29, 30] and mould shrinkage can be lessened [35].

8.3 Reinforcement using LCPs

Fibre-reinforcement has been used for many years to increase the engineering performance of thermoplastics. Such composite materials have found applications in the aerospace, automotive and marine industries among others. The reinforcing materials are traditionally inorganic fibres such as graphite, boron and glass. A more recent development has been the use of organic materials such as polyaramids, e.g. Kevlar, to give lightweight composites with improved properties. Organic fibres are beginning to displace inorganic reinforcing materials in some application areas. Kevlar itself is a liquid crystal polymer and is solution spun in its lyotropic state to produce fibres. It cannot be melt spun because it decomposes above its melting point. Composites containing Kevlar are therefore examples of liquid crystal blends in commercial use.

In recent years thermotropic liquid crystal polymers have become available and can be processed in the melt to give highly oriented structures that are largely retained on cooling and subsequent crystallisation. The rod-like molecular conformation and chain stiffness give LCPs their much vaunted 'self-reinforcing' properties that are close to those of fibre reinforced composites. Figure 8.11 illustrates the effect of reducing the level of aromaticity and chain stiffness on tensile modulus.

There are a number of thermotropic melt-spun fibres available, the most notable of which is Vectran (available from Hoechst-Celanese). Heat treated Vectran fibre is reported to have a tensile strength about equal to that of Kevlar 49 at room temperature [37].

The use of solid fibres in thermoplastics is not without its problems. Heavy loadings cause a substantial increase in melt viscosity and generally lower the ease of processing. Increases in viscosity may lead to problems such as machinery exceeding the operating limits, resulting in flash or short shots, freezing of thin-walled sections, etc. Raising the temperature often leads to thermal degradation and increased energy expenditure. An additional difficulty is the wear of extruders/fabricators due to abrasion (e.g. with glass reinforced plastic) that can cause contamination by metal fragments and change in critical dimensions in mouldings and dies. Such abrasion can cause drift in process product quality even before significant damage is apparent.

The generation of reinforcing species *in situ* offers advantages over the addition of solid fibres and fillers. The use of techniques such as *in situ* crystallisation and polymerisation have been reviewed by Kiss *et al.* [38] and have shown limited success. Reinforcement through the addition of LCPs to thermoplastics has been investigated over the last few years with some encouraging results. The principal goal is to achieve improvements in mechanical properties by using the LCP component to reinforce the flexible thermoplastics through the formation of fibres. Blending is also considered as a possible route to overcome the highly anisotropic physical properties of LCPs that can be problematic in many applications. The LCP producers were

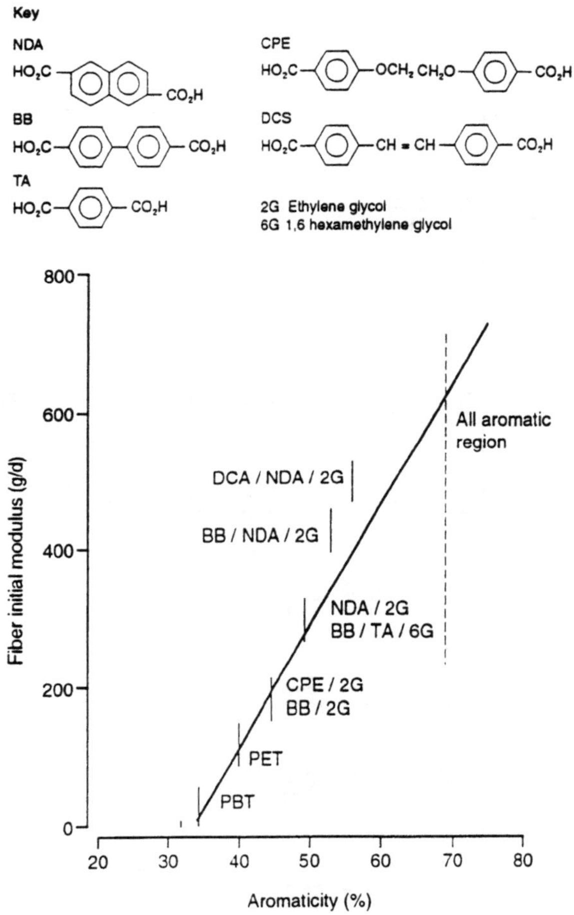

Figure 8.11 Effect of polyester aromaticity of fibre tensile modulus (after Calundann and Jaffe [36]).

among the first to publish studies in this area and showed that many of the desirable properties of the pure LCP material can be sustained even when it is the minor component of the blend [34, 39]. The challenge is to produce organic LCP fibres *in situ* that give the required level of reinforcement at a realistic cost.

The use of LCPs in polymer blends looks attractive from a number of view points. First, their high para-linked aromatic content gives polymers with good moduli. Second, LCPs have very low melt viscosities allowing good flow properties and ease of processing. They readily form fibrous structures, injection moulded samples of neat LCP resin having an appearance similar to wood. Other useful properties include high chemical resistance and low mould shrinkage (useful for accurate mouldings).

Table 8.4 Mechanical properties of *in situ* composites from recent literature. Two- or three-dimensional samples are injection or compression moulded, one-dimensional samples are extruded strands, spun fibres or extruded sheets. Properties are for an LCP content of 25 wt%, unless indicated otherwise. (After Crevecoeur and Groeninckx (1990) *Bull. Soc. Chim. Bela.* **99**, 1031–1044 [41b]).

		Two-, three-dimensional						One-dimensional					
Matrix	TLCP	E (GPA)	Δ (%)	σ (MPa)	Δ (%)	ε (%)	Δ (%)	E (GPA)	Δ (%)	σ (MPa)	Δ (%)	ε (%)	Δ (%)
PA	Vectra B							1.1	116	65	28		
	Vectra A	4.4	83	120	40	12	−88						
	Vectra A							6	200	100	335		
	Other							2.0	132	69.7	31	75	−56
	Vectra A	2.65	29	66.9	−8	14.6	−97						
	Vectra B	3.16	53	60.7	−17	2.5	−99						
PC	Vectra A	5.72	147	121	81	3.49	−97						
	Vectra B	6.55	182	154	130	4.2	−96						
	Vectra A							1.4	−13				
	LCP 2000	1.4	40					3.2	68				
	Other							7.09	198	342	246	6.9	−92
	Other	0.8	11					7.1	223				
	PET/PHB							4.1	46				
PET	PET/HB	1.17	23	43.7	9			2.54	32	67.8	12		
	LCP/2000							0.2	111				
	PET/PHB	4	80										
	Other							6.97	347	294	332	4.4	−98
	PET/PHB							7.5	416	60	31	2.2	−99
PS	PET/PHB	1.67	15	22.5	−20			2.63	29	37.5	−14		
	Other							3.59	50	95	20		
	Vectra A	3.8	19	33.4	4	1.04	−1	15.0		135		1	
	Vectra B	4.8	50	38.7	21	0.89	−15						
PBT	Vectra A	3.54	37	37.9	−27	1.2	−79						

PEI	Vectra A	5.15	69	129	42	4.3	−93						
	Vectra B	7.45	144	95.8	5	1.54	−97						
	Vectra A	8.7	53	118	3	1.4	−94	7.9	84	145	34		81
	Xydar	3.1	10	113	−23								
PAR	Vectra A	4.27	1.81	102	44	5.3	−97						
	Vectra B	3.28	116	51.7	−27	2.0	−99						
	Ekonol 6000	2.64	56	110	73	20	−75						
	Ekonol 2000	2.48	47	90	42	10	−88						
PEEK	Vectra A							4.3	23	71.7	−15	2.5	−95
	Vectra A	8.7	47	112	20								
PES	Vectra A	4.99	100	125.5	97	3.8	−97	5.79	123	164.1	131	3.6	−97
	Vectra B	8.82	253	172.4	171	2.6	−98	11.8	354	139.9	97	1.9	−99
PPE/PS	Vectra B	6.1	120	65.3	−8	1.3	−77	15.1	410	95	−58	0.7	−99

1. 30 wt% TLCP
2. Better mechanical properties for lower TLCP contents, e.g. 3.6 GPa for 10%
3. 20 wt% TLCP
4. 10 wt% TLCP
5. Measured at 80 °C
6. 16 wt% TLCP
7. Matrix consists of miscible blend of 70 wt% PPE and 30 wt% PS

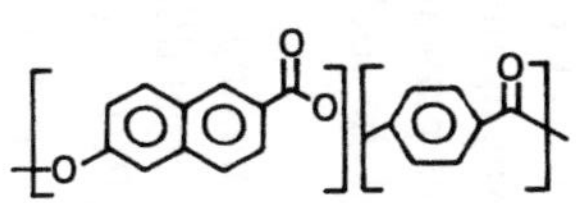
Vectra A Senes

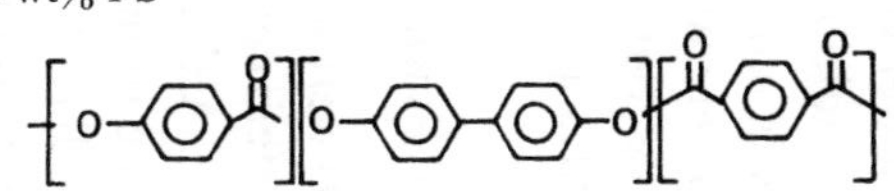
Ekonol 2000/
Xydar SRT 300

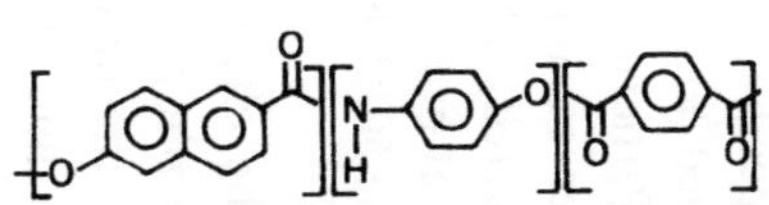
Vectra B Senes

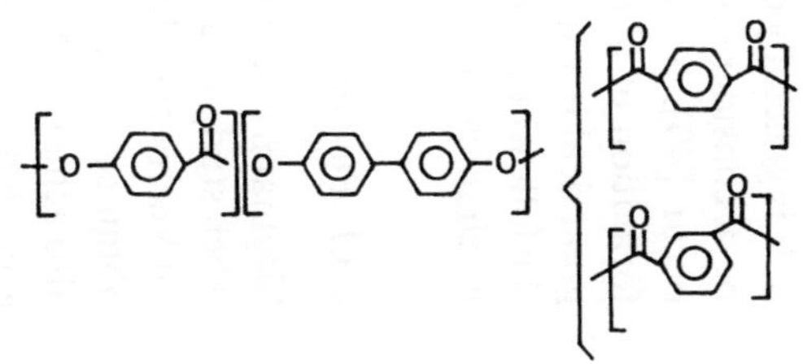
Ekonol 6000

The majority of studies to be found in the literature involve blends in which the component polymers are immiscible. Because of the limited availability of LCPs, studies have concentrated on the commercially available materials. Table 8.1 details the thermal characteristics of some of the commercial LCPs together with their formulation. Table 8.2 gives examples of their mechanical properties taken from manufacturers' data sheets. From these tables it can be seen that the polymers have properties comparable with reinforced engineering plastics such as PPS, poly(etheretherketone) (PEEK), etc. and have correspondingly high processing temperatures. Much of the work carried out so far involves LCP blends with such polymers.

It is generally considered that the viscosity of the additive component must be of considerably lower viscosity than the host matrix for fibril formation to occur [40]. This condition is valid for most LCP polymer blends so the formation of fibrils *in situ* is regarded as straightforward. There has been a great deal of experimentation in this area, some of which will be discussed in the following sections, but for convenience a summary of the improvements in mechanical properties achieved to date is given in Table 8.4.

8.3.1 Blending of LCPs with thermoplastics to improve mechanical properties

Siegmann *et al.* [19] were among the first to report *in situ* composite formation when studying a system containing amorphous polyamide and an experimental Vectra type LCP. Although the work concentrated on the positive results obtained through the improvements in processing, some reinforcement was also achieved. A fibrillar structure was observed when the LCP concentration exceeded 25%. Blend morphology was characterised by scanning electron microscopy of etched samples. Evidence was found for a higher degree of fibril formation at the skin rather than at the core of injection moulded specimens. There was some improvement in elastic modulus (Figure 8.12) and tensile strength, but ultimate elongation sharply decreased (Figure 8.13).

Similar results were reported by Chung [41a] using nylon 12 and an unspecified HNA/HBA (Vectra type) LCP. The extruded blend again displayed an anomalous skin–core structure that is a recurrent theme in LCP/thermoplastic blends. Improvements in mechanical properties were observed until the LCP concentration exceeded 80% (phase segregation occurred). This was again achieved at the expense of a sharp drop in elongation at break. This plateaued out with increasing LCP content. Interestingly, in the 50/50 blend the coefficient of thermal expansion was found to approach zero.

The effect of mixing Vectra A and B grades with a wide variety of amorphous and crystalline polymers including polyether sulphone (PES), polyetherimide (PEI), polyarylate, polyacetal, nylon 6, polybutylene terephthalate (PBT), polycarbonate (PC), poly(chloro-trifluoroethylene) and poly(etheretherketone) was investigated by Kiss [39]. The blends were pro-

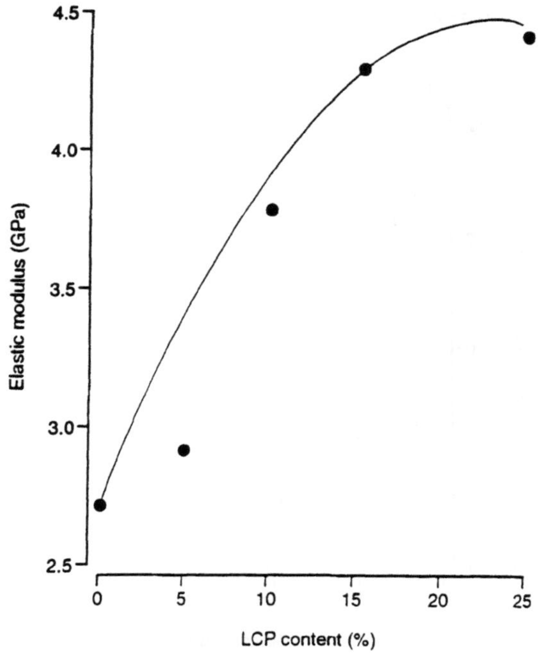

Figure 8.12 Composition dependence of the elastic modulus of injection moulded LCP/PA blends (after Siegmann *et al.* [19]).

duced by a number of compounding techniques and either extruded into strands or injection moulded. No distinctions were made between the varying compounding methods and the exact processing conditions were not cited so it is difficult to interpret any structure–processing–property relationships. However, a general trend was observed: dramatic increases in the tensile and flexural moduli (two to three times), with smaller increases in flexural and tensile strength, accompanied by a substantial decrease in elongation to break. The morphologies of the blends were found to be similar to those reported in other studies, with elongated LCP domains that ranged from extended droplet to large fibres. No mention was made of any skin–core morphology but differences in adhesion of the LCP phase in the various systems were apparent.

The extent of property improvement is dependent on the method of compounding. This effect was demonstrated by Swaminathan and Isayev [42] who prepared blends of PEI and Vectra A950 using a static mixer and a co-rotating twin-screw extruder. They found that the static mixer gave blends with a higher degree of fibrillation and corresponding improvements in Young's modulus over those produced using the twin-screw extruder. SEM was used to show that the LCP phase existed as droplets or elongated

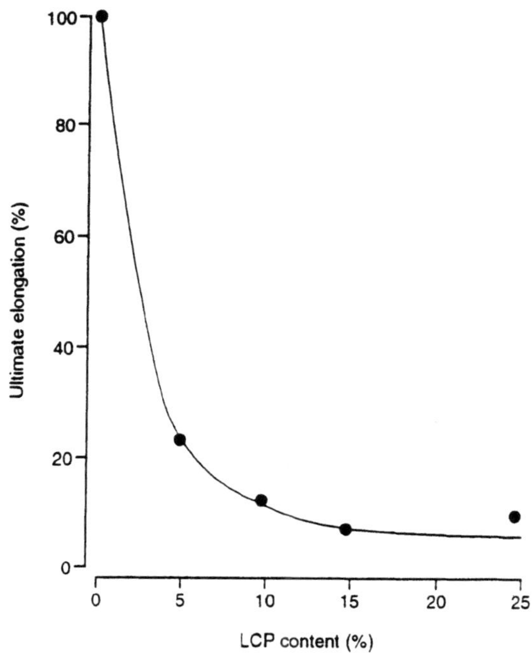

Figure 8.13 Composition dependence of the ultimate elongation of injection moulded LCP/PA blends (after Siegmann *et al.* [19]).

structures at concentrations below 10% and only formed fibrils (typically 2–5 μm diameter) when the LCP component was greater than 20%. It was reported that the LCP fibres formed at the inlet zone of the die where the polymer experienced high elongational stresses. Further, the authors reported that the modulus and strength of the extrudates from capillary die with shorter *L/D* ratios were higher than those with longer *L/D* ratios. The interphase adhesion between the PEI and the LCP was found to be poor and skin–core morphology was apparent in injection moulded specimens.

Blizard and Baird [18] have studied blends of PET/HBA (40:60) with nylon 6,6 and polycarbonate with the aim of determining the factors effecting fibril formation. SEM was used to assess the extent of fibrillation in the samples prepared by extrusion through capillary dies of various *L/D* ratios and immediate quenching. Fibrils were only evident when the LCP concentration exceeded 30%. In a simple cone and plate rheometer fibrils could not be produced, even in the blends of higher LCP content (> 30%) and shear rates that ranged up to $100\,s^{-1}$. Capillary rheometry showed that fibrils are readily produced with a die of $L/D = 7.82$ but not with an $L/D = 21.4$. This suggested that although it may be possible to produce fibrils using simple shear flow, extensional forces readily lead to fibril formation. The authors concluded that processing through capillary dies with a short land results in articles with higher mechanical properties. This view is substantiated by the work of La Mantia *et al.* [24] with their study of Vectra B950 and nylon 6. SEM analysis of a 20% LCP blend extruded at a shear rate of $1200\,s^{-1}$ through a capillary die ($L/D = 40$) showed no fibrillation. In contrast, if the blend is passed through a die of $L/D = 0$, with conical inlet under identical conditions, fibrils are produced. Thus fibrils can be lost during the flow in a long capillary if, for example, the orientational relaxation time of the liquid crystal polymer is less than the time necessary to flow in the capillary.

Extensional flow is inherent in film production and was the method chosen by Ramanathan *et al.* [43] to study blends of PC with Vectra A900 and PET/HBA (40:60). The authors reported doubling of both tensile strength and modulus in the range of the draw ratio used for the PET/HBA blend. No evidence of fibril formation or improvement was noted for the Vectra-based blend and this was attributed to the higher processing temperature used. At 260 °C, the viscosity of the LCP component is likely to be higher than the polycarbonate. In a further publication, Ramanathan *et al.* [44] examined blends of Vectra A900, PET/HBA (40:60), PET/HBA (20:80) and a 50/50 blend of the two PET/HBA materials with polyphenylene sulphide. Extruded film with a draw ratio of up to 9 showed fibril formation in both the PET/HBA formulations containing PET/HBA (40:60) but not in the other two cases. They attributed the lack of fibrillation in the PPS/Vectra A900 system to possible chemical reaction between the two components under the prevailing processing conditions.

In a more recent paper Lee and Dibenedetto [45] suggested that such chemical reactions can be used to improve the adhesion between incompatible aromatic fibres and the thermoplastic matrix. In a preliminary study, LCP/LCP composites based on KU-9211 (Bayer AG) and PET/HBA (40:60) have been prepared in which the KU-9211 acts as the reinforcing phase. Characterisation of the extruded and drawn samples showed that there are significant chemical interactions between the two thermotropic LCPs. The next stage is to introduce a third host matrix. This appears to be an interesting approach given that PET/HBA (40:60) has been reported to bond well with conventional thermoplastics such as PET [46] and PHMT [47].

The effects of miscibility of the host and reinforcing phase on the development of *in situ* composites have been studied by Zhang and co-workers [48]. Solution and melt blends of PET/HBA (40:60) with PS, PC and PET were examined using DSC, SEM and dielectric thermal analysis (DETA). It was established that the LCP was immiscible with PS while there was some degree of miscibility in the PC and PET cases. The gradation in LCP interaction was evident in the morphology of the prepared blends, with the PS composite having the most coarse and defined morphology; the least distinctive, finest morphology was associated with the more miscible PET system. The transition from ellipsoidal to fibrous LCP domains was found to occur with increasing shear rate in capillary extrudates as well as meltspun fibre for all three blends.

Compatibility between PET and PET/HBA (40:60) has been investigated by Brostow *et al.* [49]. They reasoned that there was a limited degree of solubility of the LCP in the PET associated with the PET-rich phase of the LCP component. Mechanical properties were observed to decrease initially with the addition of a small amount of the LCP (2–5%), then increase to a maximum and fall off again as the LCP content increased. Friedrich *et al.* [50] studied the same system and blends of PC and PET/HBA. The DSC results with PC/LCP blends showed a single T_g on annealing which was interpreted as partial miscibility through transesterification. In contrast, the two T_gs observed in PET/LCP blends (the lower associated with the PET-rich phase of the LCP and the higher with the HBA-rich) were unaffected, indicating no miscibility. Using polarising optical microscopy, however, the birefringence emanating from domains of LCP decreased on annealing indicating a loss of liquid crystal order through transesterification. It is important to note that although transesterification appears to take place in both systems studied the time-scale of reaction is far longer than any typical processing time.

Joseph *et al.* [46] used DSC to show that PET/HBA (40:60) enhanced the rate of crystallisation of PET and suggesting a possible nucleating effect. This has been supported by the work of Bhattacharya *et al.* [51] who also observed both increased nucleation and significant orientation of the crystallites in melt spun PET/LCP blends. Similarly, WAXD was used by Sharma *et al.* [52] to show that significant levels of crystalline orientation can be produced in PET/LCP blends during extrusion under conditions that produce no orientation in PET.

In a recent study by Nobile *et al.* [23], the possibility of transesterification was explored using FTIR. Blends of PC and PET/HBA were compounded using a single screw extruder and pelletised. The PC component was solvent extracted and compared with the blend and pure PC using FTIR for any evidence of ester-exchange with the LCP. No absorption bands were found at characteristic ester group frequencies and inherent viscosity measurements of the extracted PC showed no significant change from the unfilled control. On the basis of these results the authors concluded that no transesterification had occurred on processing of the blends. Jung and Kim [53] confirmed the earlier work of Friedrich *et al.* [50] showing that the T_g of PC is decreased by a small addition of PET/HBA (40:60) although the glass transitions of the LCP remain unchanged. This was interpreted as possible exclusion of PC from the LCP phase while there was some partial mixing in the PC phase. This study also underlined the importance of extensional flow on fibril formation; strands of blend containing 30% LCP produced through capillary rheometry showed no evidence of fibril formation although a 5% LCP blend that had a draw ratio of 15 gave an extended fibril morphology. Good interphase adhesion was found between the two polymers.

Changes in crystallisation behaviour have been reported in other systems. For example, Minkova *et al.* [54] found that an increase in temperature of non-isothermal crystallisation was achieved when Vectra B was added to PPS. The increased nucleation density of the blend was found to be independent of concentration. Shin and Chung [55] showed that PET was nucleated by an unusual experimental LCP (m.p.t. 272 °C) containing long methylene spacer units. The blend components were found to be immiscible but the strong interphase adhesion between the LCP fibrils and the PET resulted in good fibre reinforcement. By contrast, Pracella *et al.* [56] showed that the rate of crystallisation of PBT was reduced when blended with liquid crystalline poly(decamethylene 4,4′-terephthaloyl dioxydibenzoate).

8.3.2 Application of composite theory to in situ *composite reinforcement*

Composite theory has been used by a number of authors to describe the mechanical properties of LCP reinforced blends. Crevecoeur and Groeninckx [57] used the Halpin–Tsai equations [58] to estimate the modulus of Vectra B950 fibres within an LCP/PPE-PS (70:30) blend. The theory describes the modulus of a composite material composed of stiff unidirectional fibres as a function of fibre content and aspect ratio. Furthermore, a perfect mathematical interface between the uniformed fibres and the matrix is assumed.

$$E_c = E_m \cdot (1 + \tau\eta V_f)/(1 - \eta V_f)$$

where

$$\eta = (E_f/E_m - 1)/(E_f/E_m + \tau)$$

and

$$\tau = 2(L/D)$$

E_c = composite modulus in the fibre direction, E_m = modulus of the matrix, E_f = modulus of the fibre and V_f = volume fraction of the fibre.

The equation simplifies to the well known ‘rule of mixtures’ for the modulus for infinite fibre aspect ratio (L/D). Very large L/D ratios are not uncommon in many of the LCP blends studied to date although good interphase adhesion is unusual.

$$E_c = V_f E_f + V_m E_m$$

where V_m = volume fraction of matrix. The applicability of this approach can be assessed using a plot of volume fraction of LCP fibre versus tensile modulus (Figure 8.14). The linear relationship indicates that the blend under consideration must contain fibrils of high aspect ratio and a near perfect level of orientation. In general, blends prepared as melt spun fibre appear to be suited to this treatment despite reservations about the effectiveness of bonding between the two components. Figure 8.15 illustrates the importance of aspect

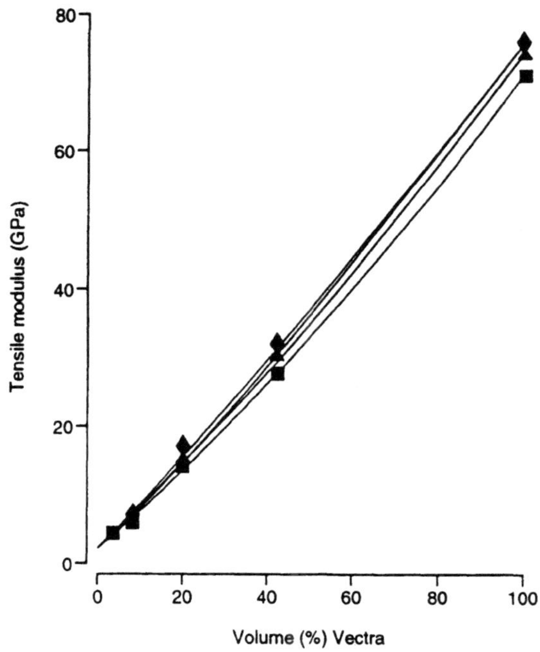

Figure 8.14 Tensile modulus versus volume fraction Vectra for PPE/PS (70/30)-Vectra B950 fibres. Draw ratio = 10; ■ DR = 20; ▲ DR = 30; ◆ DR = 40 (after Crevecoeur and Groeninckx [57]).

ratio on modulus. Sharp deviations from the rule of mixture's relationship are evident only when the fibre aspect ratio falls substantially below 100.

Blizard *et al.* [59] studied blends of Vectra A950 and an LCP supplied by Granmont Inc. (a random polyester of terephthalic acid, (1-phenylethyl) hydroquinone, and phenylhydroquinone) with PC and PEI. Using the simplified Halpin–Tsai equation, the results of fibre spinning experiments were used to estimate the tensile modulus for the two LCPs. They found an average modulus of 24.6 GPa for Vectra A950 and 23.3 GPa for the Granmont LCP. The calculated modulus for Vectra fibrils in the composite is comparable to the value for the LCP when extruded neat with similar draw ratio. However, it is considerably lower than the literature values of melt spun Vectra moduli of 41 GPa at a draw ratio of 6 and 62 GPa at a draw ratio greater than 50. The authors concluded that there was considerable potential for further improvement in the blends through processing modification.

Shin *et al.* [60] used the modified Halpin–Tsai equation to predict the aspect ratio of the fibres produced by extrusion and drawing of blends of PET and two semi-flexible thermotropic polyesters. They found that in the case of the more aromatic polymer, an aspect ratio of > 50 was readily achievable and

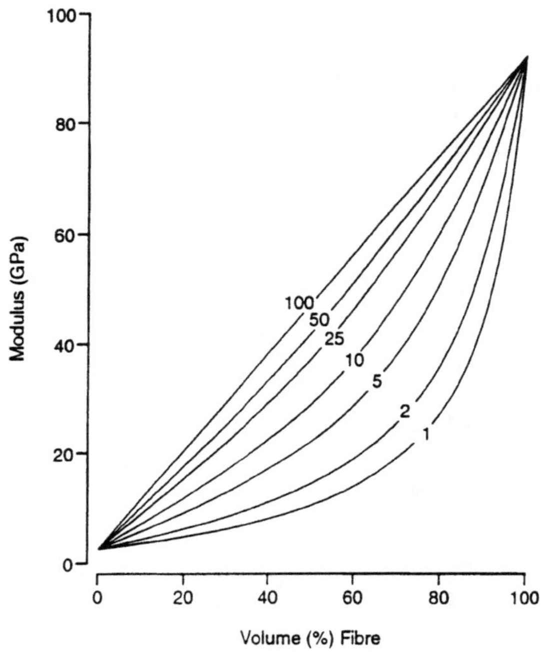

Figure 8.15 Graphical representation of the Tsai–Halpin equation with $E_m = 2.5$ GPa and $E_f = 75$ GPa. Aspect ratios indicated in the figure (after Crevecoeur and Groeninckx [57]).

gave a three- to four-fold increase in tensile modulus. In contrast, only limited reinforcement could be produced with the lower melting, less aromatic LCP.

A procedure for estimating the ultimate strength of *in situ* composites has been described by Crevecoeur and Groeninckx [41b, 57]. This approach examines the two limiting cases of Halpin–Tsai theory. First, if the strain to break of the matrix material is much higher than the LCP, at low LCP concentration the ductile failure at high strain is dominated by the matrix and will determine the strength according to:

$$\sigma_c = (1 - V_f)\sigma_m$$

where σ_c = ultimate strength of composite and σ_m = ultimate strength of matrix.

Second, at high LCP content, the composite will be dominated by brittle failure at low strain and the ultimate properties can be described by:

$$\sigma_c = V_f \sigma_f + V_m \sigma_f \cdot E_m/E_f$$

where σ_f = ultimate strength of fibre.

The experimental data for PS/PPE-Vectra B950 blends were found to be lower than theoretically predicted (Figure 8.16). The authors explain that a

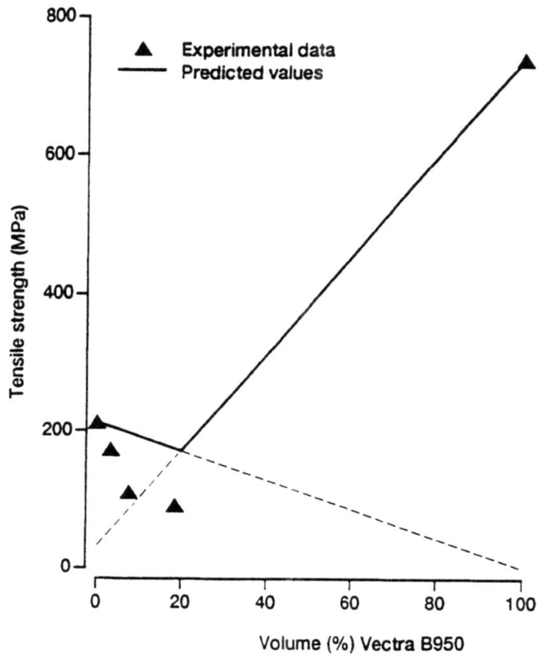

Figure 8.16 Tensile strength versus composition for PPE/PE-Vectra B950 composite fibres (DR = 30) (after Crevecoeur and Groeninckx [41b]).

limitation of this approach is that non-linear stress–strain behaviour at high deformation is not accounted for. Nevertheless, the shape of the behaviour was qualitatively predicted and the method is useful for assessing the reinforcing potential of particular LCP-blend combinations.

8.3.3 Novel methods of processing LCP/thermoplastic blends

As demonstrated by the above discussion, the majority of the LCPs used in blending are wholly aromatic, high melting materials primarily designed for engineering applications. This limits the range of matrix polymers that can be reinforced as: (i) the processing windows are required to overlap and, generally, (ii) the viscosity ratio of the LCP and the blend need to be low [40] for effective *in situ* fibril formation. The exception is the semi-aromatic PET/HBA series that has been blended with resins such as PS [61] and PVC [29, 30] but only moderate improvements have been obtained despite the formation of extended fibrillar morphologies.

The problem with producing a lower melting LCP is that the mechanical properties of the polymer are compromised. Figure 8.11 illustrated how the level of aromatically affects the modulus. However, LCPs containing aliphatic units do have some advantages, e.g. their ability to blend with lower melting polymers such as polypropylene [62], and their improved interphase adhesion [63].

The alternative is to overcome some of the inherent processing limitations of commercial LCPs. To this end a number of novel processing techniques are being developed.

8.3.3.1 Dual extruder. An alternative to custom synthesising lower melting LCPs for blends has recently been described by Sukhadia *et al.* [64]. They use a novel processing approach that allows blends of polymers with no overlapping processing window to be prepared. In the dual extruder method, as it has been coined, the matrix and liquid crystal polymers are plasticated in two separate extruders and mixed downstream in a static mixer. The blend is finally passed through an appropriate die to give a polymer strand or sheet. The morphology developed is highly dependent on the supercooling characteristics of the higher melting LCP but fibrillation has been observed in PP/Vectra A blends [65], where the difference in normal processing temperatures of the component polymers is in excess of 100 °C. Strands of PP/LCP and PET/LCP showed significant improvement in tensile modulus over the base resins. Enhancements of 10–20 times that of the matrix material were reported with blends containing 20–30% LCP when produced by the dual extruder mixing method. These results give an indication of the property improvements possible using LCPs that have a high level of aromaticity.

Baird [66] foresees many applications for this composite production method, these include automotive panels and aircraft propellers. A similar approach is being patented by Frederico *et al.* [67].

8.3.3.2 Multi-live feed moulding (MLFM). The ease of molecular alignment in flow fields is a characteristic of liquid crystals and generally leads to highly anisotropic articles with poor transverse properties. An injection moulding technique developed by Bevis and Allan [68] at Brunel University has been used to reduce the anisotropic nature of LCP mouldings. The multi-live feed moulding process is based on the controlled shearing of the polymer melt in the mould cavity both before and during solidification. The technique involves splitting the single feed of an injection moulding machine into several ports each capable of applying pressure.

The solidifying melt can be subjected to shear, to oscillating packing pressure or to static stacking pressure. MLFM has been successfully used in moulding of pure LCP to limit the effect of weld-line defects. A very similar technique has been applied by a German company, Klocckner Ferromatik Desma (KFD, 'GTS' process introduced in 1989) in overcoming weld-line defects allowing the production of large window frames for the Airbus A340 using a commercially available LCP grade.

MLFM has also been applied to LCP blends where the shearing of the melt

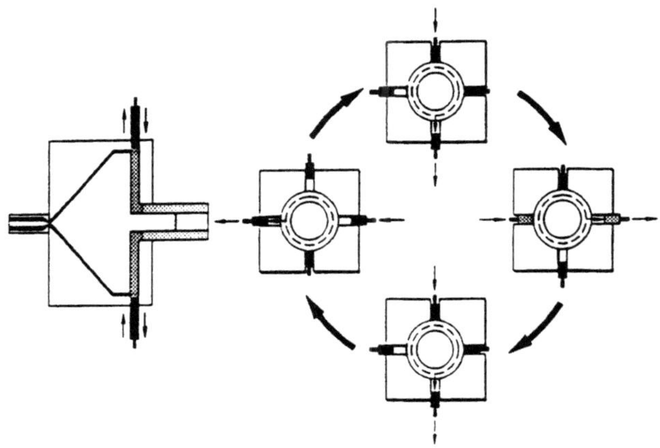

Figure 8.17 Schematic representation of the multi-live feed extrusion technique developed at Brunel University.

has produced *in situ* fibres. By using orthogonal pairs of feed gates a laminated structure can be developed. It can also be used for fibre orientation management in filled thermoplastics.

An extrusion version is also under development (Figure 8.17) and it is claimed that this will reduce defects such as voids, spider lines, etc., and greatly enhance hoop strength in applications such as pipe extrusion [69]. This type of technology is particularly well suited to the flow induced properties of LCPs and their blends which may eventually allow complex custom morphologies to be developed *in situ*.

Approaches such as these show that the problem of how best to utilise these materials is beginning to be addressed and this may encourage their wider market acceptance.

8.4 Other possible applications of LCP blends

Intrinsically LCPs offer good barrier properties to solvents and to some gases. Indeed it is very difficult to find solvents for the highly aromatic polymers such as Vectra. This is a property that has been largely ignored by blend technologists but offers considerable potential. LCPs readily elongate in uniaxial flow fields to give fibres but their biaxial orientation has been little studied. It may be possible to produce platelet domains by tape extrusion or biaxial processing methods such as film blowing. If such structures could be achieved they should lead to improved barrier resistance. Blizard [70], with the technology development company Foster-Miller, has recently started to investigate this area and has produced biaxially oriented blown films of LCP blends (and pure LCPs). Suggested end uses for such films include: substrates for electronic connector components, reinforced medical tubing and materials for packaging (as a result of the high oxygen resistance of the LCP).

LCPs may have applications as blend components in membrane separations. Platé [71] suggested that blends of HDPE and an unusual LCP, tri-fluoroethoxy-phosphazene, could be used to separate gases. Kajiyama *et al.* [72] have studied the transport properties of composites containing a lower molar mass liquid crystal. They reported enhanced oxygen permeability near the crystal-nematic transition.

Huhn *et al.* [73] investigated the effect of adding low molar mass liquid crystals to PS as a plasticiser. They found that terephthal-bis-4-n-butylaniline (TBBA) was miscible with PS over a large temperature range at concentrations up to 11%. Apicella *et al.* [35] studied the dimensional stability of blends of PS with four LC materials (including TBBA) obtained as sheets by extrusion and hot-drawing. Their results suggested that improvements in dimensional stability could be obtained by adding small amounts (*c.* 5%) of polymeric liquid crystal but only if the LC component was immiscible with the PS. The fibrils produced in the fabrication process acted to constrain the surrounding polymer and gave highly oriented sheets. If a sample is exposed to high temperature then stress relaxation is observed rather than creep. This effect disappears beyond the melting temperature of the LCPs. Similarly, Zaldua *et al.* [17] showed that small amounts of X7G improved the dimensional stability of PBT extruded filaments.

Other properties of LCPs that may have applications in blends include: decreased die swell (and negative die swell), increased mould coverage, less degradation due to reduction in processing temperatures, higher continuous use temperature and lower water adsorption.

8.5 Future applications

LCPs can often reduce the viscosity of bulk polymers significantly. However, this effect has not yet been exploited commercially. There is currently insufficient understanding of the underlying rheology for LCP blends to be used commercially today. It is an area of current research interest and it is anticipated that further advances in basic science and engineering will lead to the exploitation of LCPs to modify the processing properties of a wide range of polymers in the future.

Moduli similar to those obtainable through short glass fibre reinforcement have been achieved in a whole range of thermoplastics. The tensile properties of the highly aromatic LCPs (e.g. Vectra and Xydar) offer considerable scope for further improvement but the poor interphase adhesion and morphological control of the blends has so far meant that the full reinforcing potential of the LCP has not yet been realised.

New processing techniques have shown that polymers with no obvious overlapping processing window can be melt blended by taking advantage of the supercooling characteristics of the higher melting (LCP) component. The problems of uniaxial orientation leading to poor transverse properties can be addressed using the multi-live feed approach. The process can also be used to fabricate articles where the existence of a weld line would seriously compromise the effectiveness of a component. A recent example of this is the window frames for the airbus A340 which allow a substantial weight saving to be achieved by applying LCP technology. The study of LCP/polymer blends is still in its infancy but there are a number of promising application areas.

Acknowledgement

The authors would like to thank Dr C.K. Chai for useful discussions and British Petroleum plc for their permission to publish this work.

References

1. Blumstein, A., Marget, G. and Vilasagar, S., (1981) *Macromol.* **14**, 1543–1545.
2. Percec, V. and Tsuda, Y. (1990) *Pol. Bull.* **23**, 225–232.
3. Finkelmann, H. (1987) *Angew. Chem. Int. Ed. Engl.* **26**, 816–824.
4. Gray, G.W., Harrison, K.J. and Nash, J.A. (1973) *Electron. Lett.* **9**, 130.
5. Kelker, H. and Hatz, R. (1980) *Handbook of Liquid Crystals*, Oxford University Press.
6. Friedel, (1922) *Anales Physique* **18**, 273.
7. Marchessault, R.H., Morehead, F.F. and Walter, N.M. (1959) *Nature* **184**, 632.
8. Ringsdorf, H., Schlarb, B. and Venzmer, J. (1988) *Angew. Chem. Int. Ed.* **27**, 113–158.
9. Engel *et al.* (1985) *Pure Appl. Chem.* **57**, 1009–1014.
10. Dobb, M.G. and McIntyre, J.E. (1984) *Advances in Polymer Science*, Springer, p. 61.
11. Hansson, C.B. (1972) *Physical Data Book*, Pergamon.
12. Bright, A.A. and Singer, L.S. (1979) *Carbon* **17**, 57.
13. Hedmark, P.G., Lopez, J.M.R., Westdahl, M., Werner, P., Janseen, J. and Gedde, U.W. (1988) *Polym. Eng. Sci.* **28**, 1248–1259.
14. Fox, T.G. (1965) *J. Polym. Sci. Cryst.* **9**, 35.
15. Clark, M.G., Harison, K.J. and Raynes, E.P. (1980) *Phys. Technol.* **11**, 232.
16. Doi, M. and Edwards, S.F. (1978) *J. Chem. Soc. Faraday Trans. II* **74**, 560.
17. Zaldua, A., Munoz, E., Pena, J.J. and Santamaria, A. (1991) **32**, 682–689.
18. Blizard, K.G. and Baird, D.G. (1987) *Polym. Eng. Sci* **27**, 653–662.
19. Siegmann, A., Dagan, A. and Kenig, S. (1985) *Polymer* **26**, 1325–1330.
20. Dutta, D., Fruitwala, H., Kohli and Weiss, R.A., (1990) *Polym. Eng. Sci.* **30**, 1005–1018.
21. Weiss, R.A., Huh, W. and Nicolais, L. (1987) *Polym. Eng. Sci.* **27**, 684.
22. Kulichikhin, V.G., Vasil'eva, O.V., Litinov, I.A., Antopov, E.M., Parsamyan, I.L. and Plate, N.A. (1991) *J. Appl. Polym. Sci.* **42**, 363–372.
23. Nobile, M.R., Amendola, E., Nicolais, L., Acierno, D. and Carfarna, C. (1989) *Polym. Eng. Sci.* **29**, 244–257.
24. La Mantia, F.P., Valenza, A., Paci, M. and Magagnini, P.L. (1990) *Polym. Eng. Sci.* **30**(1), 22–29.
25. Schowalter, W.R., Chaffey, C.E. and Brenner, H. (1963) *J. Colloid Interface Sci.* **26**, 152.
26. Utracki, L.A. (1989) *Polymer Alloys and Blends*, Hanser, 192.
27. Krasnikova, N.P., Dreval, V.E., Kotova, E.V. and Pelzbauer, Z. (1981) *Vysokomol. Soed.* **B23**, 378.
28. Lees, C. (1900) *Proc. Phys. Soc.* **17**, 460.
29. Lee, B. (1988a) *Polym. Eng. Sci.* **17**, 1107–1114.
30. Lee, B. (1988b) *Antec '88*, 1088–1091.

31. Delleuse, M.T. and Jaffe, M. (1988) *Mol. Cryst. Liq. Cryst.* **157**, 535.
32. Berry, D., Kenig, S., Siegmann, A. and Narkis, M. (1992) *Polym. Eng. Sci.* **32**, 14–19.
33. Toshikazu, K. (1991) PhD Thesis, University of Lowell, D. Eng.
34. Cogswell, F.N., Griffin, B.P., Rose, J.B. (1984) US Patent 4433083.
35. Apicella, A., Iannelli, P., Nicodemo, L., Nicolais, L., Roviello, A. and Sirigu, A. (1986) *Polym. Eng. Sci.* **26**, 600–604.
36. Calundann and Jaffe, M. (1982) *Proc. Robert A. Welch Foundation Conf. on Chemical Research (XXVI): Synthetic Polymers*, p. 254.
37. Brady, R.L. and Porter, R.S. (1990) *J. Thermoplastic Composite Materials*, **3**, 253–261.
38. Kiss, G., Kovacs, A.J. and Wittmann, J.C. (1981) *J. Appl. Polym. Sci.* **26**, 2665.
39. Kiss, G. (1987) *Polym. Eng. Sci.* **27**, 410–423.
40. Min, K., White, J.L. and Fellers, J.F. (1984) *Polym. Sci. Eng.* **24**, 1327.
41. (a) Chung, T.-S. (1987) *Plastics Engineering*, October, **87**, 39–41.
(b) Crevecoeur, G. and Groeninckx, G. (1990b) *Bull. Soc. Chim. Belg.* **99**, 1031–1044.
42. Swaminathan, S. and Isayev, A.I. (1987) *Polym. Mater. Sci. Eng.* **57**, 330–333.
43. Ramanathan, R., Blizard, K.G. and Baird, D.G. (1987) *SPE 45th ANTEC. Conf. Proc.* 1399.
44. Ramanathan, R., Blizard, K.G. and Baird, D.G. (1988) *SPE 46th ANTEC. Conf. Proc.* 1123.
45. Lee, W.-C. and Dibenedetto (1992) *Polym. Eng. Sci.* **32**, 400–408.
46. Joseph, E.G., Wilkes, G.L. and Baird, D.G. (1985) *Polymeric Liquid Crystals*, Blumstein, A., ed., Plenum Press, New York, pp. 197–216.
47. Croteau and Laivins (1990) *J. Appl. Polym. Sci.*, **39**, 2377.
48. Zhang, P., Kyu, T. and White, J.L. (1988) *Polym. Eng. Sci.* **28**, 1095–1106.
49. Brostow, W., Dziemianowicz, T.S., Romanski, J. and Werber, W. (1988) *Polym. Eng. Sci.* **28**, 785–795.
50. Friedrich, K., Hess, M. and Kosfeld, R. (1988) *Macromol. Chem. Macromol. Symp.* **16**, 251.
51. Bhattacharya, S.K., Tendolkar, A. and Misra, A. (1987) *Mol. Cryst. Liq. Cryst.* **153**, 501–513.
52. Sharma, S.K., Tendolkar, A. and Misra, A. (1988) *Mol. Cryst. Liq. Cryst.* **157**, 597–614.
53. Jung, S.H. and Kim, S.C. (1988) *Polymer J.* **20**, 73–81.
54. Minkova, L.I., Paci, M., Pracella, M. and Magagnini, P. (1992) *Polym. Eng. Sci.* **32**, 57–64.
55. Shin, B.Y. and Chung, I.J. (1990) *Polym. Eng. Sci.* **30**, 13–21.
56. Pracella, M., Chiellini, E., Galli, G. and Dainelli, D. (1987) *Mol. Cryst. Liq. Cryst.* **153**, 525–535.
57. Crevecoeur, G. and Groeninckx, G. (1990a) *Polym. Eng. Sci.* **30**, 532–542.
58. Halpin, J.C. and Kardos, J.L. (1976) *Polym. Eng. Sci.* **16**, 344.
59. Blizard, K.G., Federici, C., Fererico, O. and Chapoy, L.L. (1990) *Polym. Eng. Sci.* **30**, 1442–1453.
60. Shin, B.Y., Chung, I.J. and Kim, B.S. (1992) *Polym. Eng. Sci.* **32**, 73–79.
61. Amendoli, E., Carfugna, C., Nicodemo, L. and Nobile, M.R. (1989) *Macromol. Chem. Macromol. Symp.* **23**, 253.
62. Yoncheng, Y., La Mantia, F.P., Valenza, A., Citta, V., Pedretti, U. and Roggero, A. (1991) *Eur. Polym. J.* **27**, 723–727.
63. Paci, M., Liu, M., Magagnini, P.L., La Mantia, F.P. and Valenza, A. (1988) *Thermochimica Acta* **137**, 105–114.
64. Sukadia, A.M., Datta, A. and Baird, D.G. (1991a) Generation of continuous liquid crystal reinforcement in thermoplastics by a novel blending process, *NTIS AD-A224 899/5/WMS.*
65. Sukhadia, A.M. (1991b) The *in situ* generation of liquid crystalline polymer reinforcement in thermoplastics, PhD Thesis, Virginia Polytechnic.
66. Baird, D.G. (1989) taken from 'Thermoplastic-liquid crystal polymer blends'. *Advanced Composite Bulletin*, July, p. 6.
67. Frederico, C., Attalla, G. and Chapoy, L.L. (1989) European Patent Application 0 340 655 A2.
68. Bevis, M.S. and Allen, P.S. (1985) *Proc. Polymer Processing Machinery Conf. Bradford*, Paper II, p. 11/1–11/9.
69. Bevis, M.S. and Allan, P.S. (1989) UK Patent Application 89 00434.
70. Blizard, K.G. (1991) taken from 'LCP chemical structure of benefit in extrusion' *Modern Plastics International*, January, p. 14.
71. Platé, N.A. (1990) *Vysokomol. Soedin., Ser.* **A32**, 1505–1510.
72. Kajiyama, T., Washizu, S. and Takayanagi, M. (1984) *J. Appl. Polym. Sci.* **29**, 3955.
73. Huhn, W., Weiss, R.A. and Nicolais, L. (1983) *Polym. Eng. Sci.* **23**, 779–783.

9 Fibre forming blends and *in situ* fibre composites

S.C. STEADMAN

9.1 Introduction

High performance fibre reinforced composites are now becoming important engineering materials in their own right and are penetrating many market sectors previously dominated by metals. One barrier to their more widespread use is the often protracted fabrication times associated with combining the fibre reinforcement and matrix to give predictable properties in the final artifact. The use of short fibre reinforced thermoplastics, which may be processed using extrusion and injection moulding technology, has brought about significant reductions in cycle times. However, difficulties still remain in terms of controlling the fibre length distribution and orientation in finished parts, and coping with the increased viscosity of the melt as a result of incorporation of the fibres.

One promising approach to circumvent some of these difficulties is to generate the fibres *in situ*, that is, during the actual forming of the component. By careful control of the forming process it should be possible to produce fibres with, for example, an enhanced modulus.

In order to achieve this objective it is necessary to bring together two very topical yet disparate areas of research activity. The first concerns the study of the rheological properties and microstructure of extruded samples of two-component polymeric blends. This has been the subject of extensive investigation over the last ten to fifteen years (e.g. [1]) and it has been shown that under appropriate conditions the dispersed phase can deform during processing to produce elongated structures, for example, ellipsoids or even fibres. The microrheological interpretation of this has been reported by Han [2], McHugh [3] and Lyngaae-Jørgensen [4].

The second area of work relevant to the study of *in situ* fibre composites is the production of ultra-high modulus fibres. This work is discussed in section 9.3 which considers the *in situ* production of stiff polyethylene fibres from polymer blends. Polyethylene is chosen because it is a natural fibre-forming polymer and detailed knowledge exists of production techniques for the achievement of high modulus. A variety of novel techniques have been developed over the past fifteen years or so, including capillary extrusion [5], uniaxial drawing [6], hydrostatic extrusion [7] and gel spinning methods [8]. It is the first two of these approaches which are of greatest interest to the study of *in situ* fibre formation.

A brief introduction to the three main aspects of *in situ* fibre formation follows. To begin, a summary of two component polymeric blends is presented. The next two sections will introduce the relevant aspects of capillary die (ram) extrusion and uniaxial drawing, as applied to the production of high modulus polyethylene. The two methods are also discussed in the context of the present work, that is, the production of self-reinforcing polymer blends.

9.1.1 *Two-component polymeric blends*

Today there are hundreds of homopolymers commercially available. Since one type of polymer does not possess all the physical and mechanical properties desired in a finished product, it is natural to try to use two or more polymers in order to meet the requirements. When two homopolymers are mixed there are a number of fundamental questions that must be answered. Some important questions are:

(a) Will the mixture of the two polymers be compatible?
(b) How can uniform dispersion be ensured?
(c) Will the blend possess any special morphological features?

For the purpose of *in situ* fibre formation, an exploitation of the blend morphology is of paramount concern.

Blend morphology is initially dependent on the way in which the two components flow during forming. Two types of multiphase flow can be distinguished on the basis of their degree of phase separation. One type is 'dispersed' multiphase flow, in which one component exists as a discreet phase dispersed in another component forming the continuous phase. The other type is 'stratified' multiphase flow, in which two or more components form continuous phases separated from each other by continuous boundaries. In general, only dispersed multiphase flow is of relevance when applied to the *in situ* fibre formation of one polymer in a blend. In dispersed multiphase polymeric systems there are many interrelated variables which affect the ultimate physical and mechanical properties of the finished product. Such relationships are shown schematically in Figure 9.1. For instance, the method of preparation (for example, the method of mixing the polymer blend) controls the morphology (the state of dispersion, particle size and particle shape) of the mixture. On the other hand, the rheological properties strongly dictate the choice of processing conditions (for example, temperature and shear stress) which in turn strongly influence the morphology and therefore the ultimate physical and mechanical properties of the finished product.

The work described in section 9.3 is concerned with the production of a dispersed polymer blend, based on two incompatible polymers, in which one polymer forms a discrete phase within the other (continuous phase). By suitable control of the drawing process the discrete particles can be drawn to form fibres. This has been achieved using the two methods described earlier,

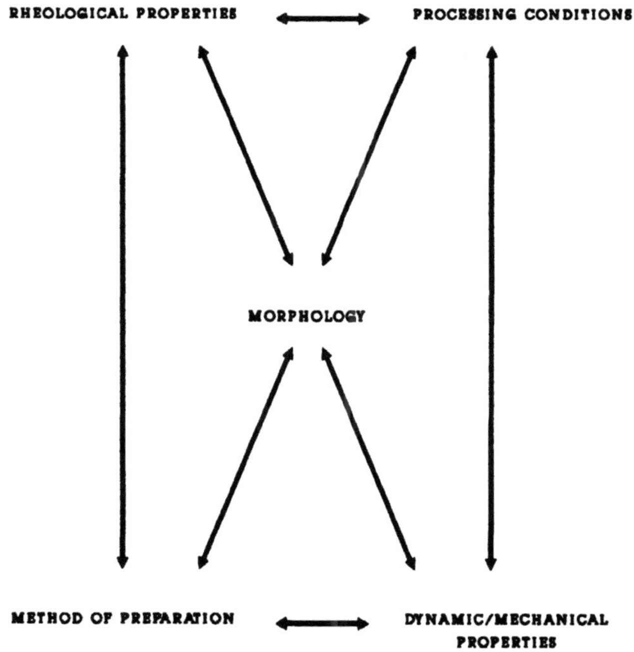

Figure 9.1 Variables affecting physical and mechanical properties of polymer blends (after Han [2], by kind permission of Academic Press).

namely capillary extrusion and uniaxial drawing. The relevant aspects of these processes are now presented.

9.1.2 Capillary die (ram) extrusion

The occurrence of orientation-induced crystallisation in flowing polymer melts was first reported by Van der Vegt *et al.* [9]. Molten polyethylene or polypropylene, when transported through a capillary die at a temperature close to the melting point, was reported to undergo a longitudinal elastic extension which led to spontaneous crystallisation of the melt. Rapid solidification of the melt caused the capillary to block, preventing further flow of the material. This was followed by the recognition that the blocked material extracted from the capillary had special properties, including a trend towards high modulus. This has been reported by Porter *et al.* [5]. Much work was subsequently conducted in this field, aimed at distinguishing between the contributions of flow-induced crystallisation and solid state deformation to the observed enhancement in modulus. More recent trends (e.g. [10]) tend to attribute the resulting high modulus primarily to the solid state extrusion process which occurs when excessive pressures are maintained.

A much simpler method of using the above principle would be to create this phenomenon in a two-phase system. When the crystallisable polymer is embedded in a matrix which does not crystallise under the extrusion conditions, the elongated crystals formed by the enclosed phase can be easily transported through the die. After complete solidification the extrudate is uniaxially reinforced by parallel fibres of high stiffness and strength.

9.1.3 Uniaxial drawing

A number of values have been proposed for the Young's modulus of fully aligned high density polyethylene (HDPE) on the basis of theoretical calculations conducted by Frank [11]. All these estimates, although far from being in good agreement among themselves, indicate that a Young's modulus of about 250 GN/m^2 should be expected for a completely oriented sample. This value is several orders of magnitude higher than that usually achieved by cold drawing conventional HDPE.

As a result of the developments of Capaccio and Ward [12], the two objectives of extremely high draw ratios and moduli have become indisputably linked. They have reported that provided certain criteria for molecular weight, initial morphology and drawing conditions are satisfied, elongations of up to 30 times can be achieved, along with moduli of up to 70 GN/m^2 (i.e. up to $\sim$ 1/4 or 1/3 of the theoretical maximum).

Ultradrawing processes can normally only be used to produce small sections of material in the form of monofilament, multifilament yarn or tape. Although the moduli obtained for HDPE are comparable with E-glass or aluminium, there still exists a requirement for the material to be incorporated into a suitable matrix prior to manufacture of the final artifact. A more convenient method would be to conduct the drawing process on a two-phase material in which one phase exists as a discrete, naturally fibre-forming polymer. Twin-screw extrusion of the two phases may be followed by a controlled drawing process at the die exit to orient and extend the dispersed fibre-forming polymer within the matrix. The final extrudate will consist of a matrix which is reinforced by parallel fibres of high stiffness and strength. Selection of a matrix material with a lower processing temperature than the melting point of the reinforcing fibres will allow the composite to be processed using compression and injection moulding technologies, without destroying the high draw ratio built into the fibre phase.

9.2 *In situ* composites: a review

It is well known that the mechanical properties of polymers can be greatly affected by the inclusion or addition of filler materials. Each of the many types of fillers used, such as elastomers for impact resistance, and mineral filler or glass beads to increase heat distortion temperature, contribute to the balance of properties required for specific applications.

In particular, the strength and modulus of a polymer can be greatly improved by the addition of fibrous fillers, the most popular being chopped glass fibre. Glass is quite inexpensive and the benefit derived can be very significant. Unfortunately, processing and fabrication of chopped glass composites present some technical difficulties. Possibly the most significant are wear on the processing equipment due to abrasion, the increase in viscosity of the molten polymer, and difficulties in compounding. Since the glass fibre remains solid during the processing operation and is extremely hard, surfaces in relative motion such as screw flights and barrels will wear and need replacement. Even before the machine is damaged, the product quality may suffer because of drift in the process, contamination by metal from wearing surfaces, and change in critical dimensions in moulds and dies.

The increased melt viscosity may lead to problems in exceeding the operating limits of the machinery, resulting in flash or short shots and 'freezing off' in thin-walled sections. Raising the temperature to reduce viscosity may lead to problems with thermal degradation. In any case, even if the part can be made satisfactorily, the expenditure of energy will be greater.

The compounding step itself can present difficulties. The filler concentration and distribution must be carefully controlled, which requires a stable compounding process and good dispersion of the fibre. The fibre length distribution can be greatly affected by breakage during compounding. If nothing else, the compounding step is an extra expense because of the cost of the machinery and the time required to perform it. For these and other reasons, it would be highly desirable to find an approach in which the reinforcing species is not actually present before the processing of the resin, but comes into existence during the processing. The *in situ* formation of the reinforcing species has led to the term *in situ* composite to describe materials of this type. A review of these materials is now presented.

9.2.1 In situ *crystallisation*

The earliest approach to the formation of *in situ* composites was the concept of *in situ* crystallisation. In this approach, an additive is introduced into the polymer, which is molten at the processing temperature, and crystallises into elongated crystals as the polymer cools. Some specific attempts are reviewed by Kiss *et al.* [13] and will not be repeated here. However, as Joseph *et al.* [14] have pointed out, there is no real hope of obtaining high performance composites by using organic crystals as a reinforcing additive, since these crystals are held together by relatively weak Van der Waals forces.

9.2.2 In situ *polymerisation*

The extension of *in situ* crystallisation is the concept of *in situ* polymerisation. As explored by Kiss *et al.* [13], the idea is to form needle-like crystals of a monomeric species during processing, and then to polymerise the monomer to obtain a highly oriented fibrous polymer. Specifically, trioxane was crystallised in a matrix of polycaprolactone or polyoxymethylene, then solid-state polymerised by gamma radiation. This work was unsuccessful for two reasons. First, the yield of the solid-state polymerisation was too low, leading to numerous voids in the final product. Second, the solid-state polymerisation of trioxane to polyoxymethylene was not a true topochemical reaction. It was observed by Wegner *et al.* [15] that only about 55% of the polyoxymethylene was found in the extruded chain form oriented in a direction parallel to the original trioxane crystal, the remainder forming a twinned structure at an angle of 76.7 °. Thus, only about half of the limited amount of polyoxymethylene formed was in the desired high-modulus extended chain form for reinforcement.

Greater success was obtained by workers, e.g. Baughman *et al.* [16], using monomers with conjugated triple bonds (e.g. di-and tri-acetylenes) which can be polymerised in the solid state via true topochemical reactions of nearly 100% yield. These form fibrillar extended-chain polymeric single crystals with negative thermal expansion coefficients and very high modulus and tensile strengths. A composite using this type of monomer and *in situ* polymerisation was patented by Baughman [17] on the basis of the control of thermal expansion coefficient made possible by this technique.

9.2.3 In situ *composites from blends with thermotropic liquid crystalline polymers*

This work deals with a different approach to the *in situ* composite and has much in common with the *in situ* fibre formation of two-component thermoplastic blends. In this work, the reinforcing species is also a thermoplastic, but a very special type of thermoplastic—a thermotropic liquid crystalline polymer (LCP). Thermotropic LCPs are long chain organic molecules, usually fully aromatic polyesters, that have sufficient chain stiffness to give extremely high strength, and yet have sufficient flexibility to melt. A number of examples of such polymers are cited in a review by Calundann and Jaffe [18].

Thermotropic polymers form fibres extremely readily, and even in injection moulded parts, the fibrous structure is immediately apparent. Molecular orientation occurs and the physical properties of moulded articles exhibit a high degree of anisotropy. This often leads to mechanical properties that may in principle approach those of glass reinforced materials. Difficulties remain, however, in controlling the degree of orientation and extension as well as the degree of fibrillation in moulded parts. It should also be noted that Kiss [19] has reported that a reduction in elongation to break and, in some cases, poorer weld-line performance results from the use of LCPs.

Thermotropic LCPs are easier to compound than, say, conventional chopped glass reinforced thermoplastics because of a reduction in melt viscosity. However, for the time being at least they remain very expensive. It is presently more economical to reinforce with glass fibre.

9.2.4 *Evidence for fibrillation in the flow of molten polymer mixtures*

Evidence of fibrillation in the flow of polymer blends is widespread in studies which are concerned with the microstructure and rheological properties of melt processed blends of polymers (e.g. [1]). In this section three of the most interesting are presented.

(a) Polyoxymethylene–copolyamide blends. Evidence of fibrillation in the flow of molten mixtures of polymers was first presented by Tsebrenko *et al.* [20] who examined the morphological state of polyoxymethylene (POM)/copolyamide (CPA) blends. The two thermodynamically incompatible polymers were blended and then extruded through a capillary die and examined at various points along the die by freezing samples in liquid nitrogen. Photomicrographs collected just past the die entrance indicated very little orientation of the dispersed POM phase within the CPA phase. However, extrudate collected at the downstream side of the die showed that fine, well-oriented fibrils of POM existed in the CPA matrix. In offering an explanation for the

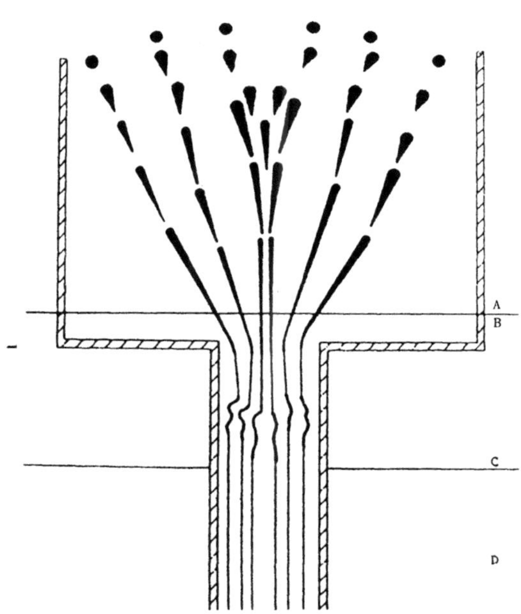

Figure 9.2 Representation of the fibrillation process in the entrance zone and in the ducts (from Tsebrenko *et al.* (1976) *Polymer* **17**, 831 [20], by permission of the publishers, Butterworth Heinemann Ltd. ©).

occurrence of the fibrillation observed experimentally, Tsebrenko *et al.* presented the schematic shown in Figure 9.2. The dispersed droplets, formed initially during the mixing operation, are elongated as they enter the die entrance (region A) under the effect of tensile stresses acting in the direction of the converging streamlines. The elongated droplets then recoil as they pass the die entrance (region C). Finally, rearrangement of the elongated droplets occurs at the downstream side of the capillary (region D), giving rise to fibrils parallel to the capillary axis. The final extrudate is therefore reinforced with uniaxially aligned fibres. The results indicate that the formation of thin fibrils takes place at the die entrance and that a prerequisite for the production of continuous fibrils is the existence of a tensile stress acting on the fibrils in the entrance zone.

(b) Polystyrene–polyethylene blends. The fibre industry also makes use of blends of two or more polymers in order to obtain certain desired mechanical and/or physical properties of a finished fibre. Using polymer blends of polystyrene (PS) and high density polyethylene (HDPE), Han and Kim [21] made an experimental study of melt spinning in order to investigate the question of spinnability and fibre morphology when two incompatible polymers are spun together. Plots of apparent elongational viscosity versus blending ratio for blends of polyethylene with two different grades of polystyrene, Styron 686 and Styron 678 are shown in Figure 9.3 and Figure 9.4 respectively. It can be seen in Figure 9.3 that the apparent elongational viscosity goes through a minimum at a blending ratio of 75/25 PS/HDPE. On the other hand, Figure 9.4 shows that as the amount of HDPE is increased in the blends of Styron 678/HDPE, the apparent elongational viscosity increases correspondingly and does not go through a minimum.

Using shear viscosity–shear rate data for the three homopolymers, Han and Kim [21] indicated that the HDPE was more viscous than the Styron 678 and that the Styron 686 was more viscous than the HDPE, i.e.

$$\text{Styron 686} > \text{HDPE} > \text{Styron 678}$$

for shear viscosity.

From this analysis it is expected that in blends of HDPE/Styron 678, the HDPE forms the discrete phase and the Styron 678 forms the continuous phase, and that in the blends of Styron 686/HDPE, the Styron 686 forms the discrete phase and the HDPE forms the continuous phase. This expectation is based on a general tendency that the more viscous component of the two forms the discrete phase and the less viscous component forms the continuous phase. It should also be noted, however, that this is also dependent on the volume fraction of the two components, the higher volume fraction component normally forming the continuous phase.

In order to explain the observed elongational flow behaviour of the two blend systems shown in Figures 9.3 and 9.4, Han and Kim [21] examined the

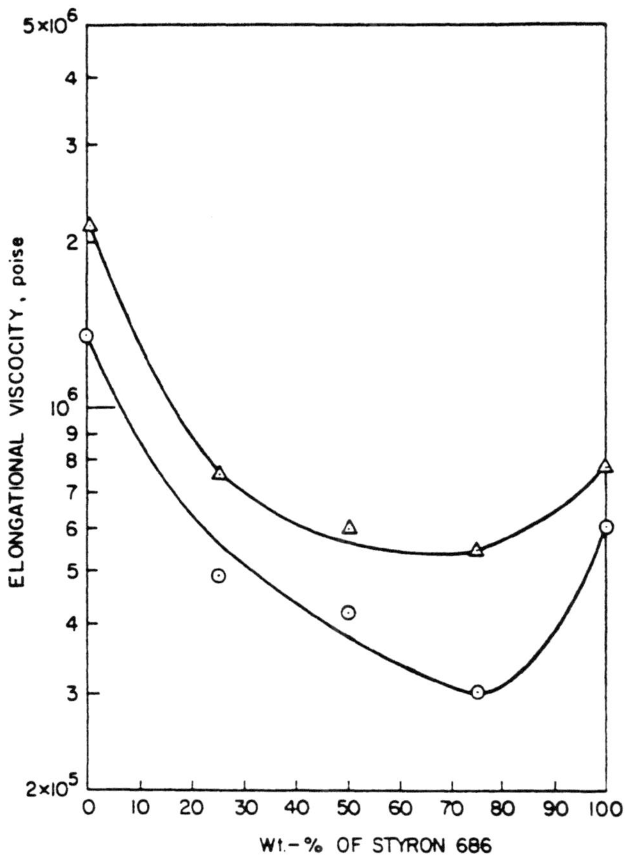

Figure 9.3 Apparent elongational viscosity versus blend composition for blends of HDPE and Styron 686, at various elongation rates as determined with fibre spinning apparatus. ⊙, $\gamma_E = 1.0\,s^{-1}$; △, $\dot{\gamma}_E = 0.5\,s^{-1}$. (From C.D. Han and Y.W. Kim, *J. Appl. Polym. Sci.*, **18**, 2589. Copyright © 1974. Reprinted by permission of John Wiley & Sons, Inc.)

elongational flow behaviour of the individual components. It was shown that the apparent elongational viscosity of Styron 686 was lower than that of HDPE, implying that in the blends of Styron 686/HDPE, at a given tensile stress the droplets (Styron 686) would be deformed more easily than the continuous phase (HDPE). On the other hand, it was shown that the apparent elongational viscosity of HDPE was higher than that of Styron 678, implying that in the blends of HDPE/Styron 678, at a given tensile stress the droplet phase (HDPE) would not deform as readily as the continuous phase (Styron 678). This observation is shown schematically in Figure 9.5. In other words, the blend system of Styron 686/HDPE shows a minimum in apparent elongational viscosity as given in Figure 9.3, due to the presence of elongated droplets of Styron 686 suspended in the molten threadline. Also, the blend

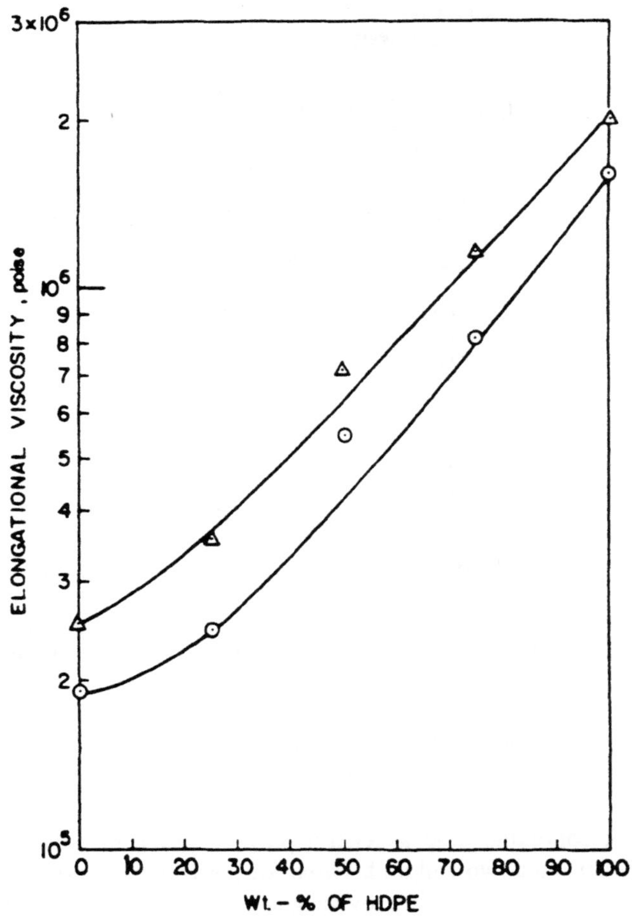

Figure 9.4 Apparent elongational viscosity versus blend composition for blends of HDPE and Styron 678, at various elongation rates as determined with fibre spinning apparatus. ⊙, $\dot{\gamma}_E = 1.0\,s^{-1}$; △, $\dot{\gamma}_E = 0.5\,s^{-1}$. (From C.D. Han and Y.W. Kim, *J. Appl. Polym. Sci.*, **18**, 2589. Copyright © 1974. Reprinted by permission of John Wiley & Sons, Inc.)

system of HDPE/Styron 678 shows a monotonically increasing trend of apparent elongational viscosity as the amount of HDPE is increased due to the presence of barely deformed droplets of HDPE suspended in the molten threadline.

(c) SBS block copolymer–polystyrene blends. Further evidence of fibre formation has been found in block copolymer–homopolymer polystyrene blends by Folkes and Reip [22] and Ehtaiatkar *et al.* [23]. This work was concerned with the effect of extrusion on the microstructure and properties of blends of a styrene–butadiene–styrene (S-B-S) block copolymer, Kraton 102,

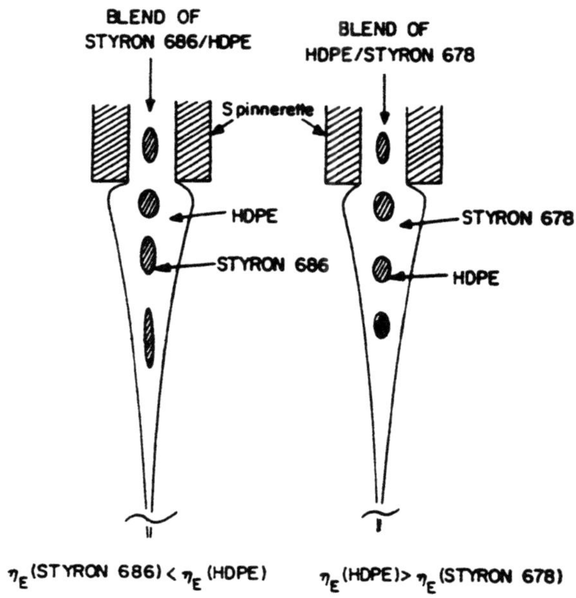

Figure 9.5 Schematic illustrating the state of dispersion and droplet deformation in the two-blend systems. (From C.D. Han and Y.W. Kim, *J. Appl. Polym. Sci.*, **18**, 2589. Copyright © 1974. Reprinted by permission of John Wiley & Sons, Inc.)

with homopolymer polystyrene. The results obtained from this investigation are discussed briefly, as they form the basis for the development of the work described in section 9.3.

The block copolymer and a general purpose grade of polystyrene were compounded using a two-roll mill. Three concentrations of polystyrene were used, namely 10%, 30% and 50% by weight. After granulation the blend was extruded using a Reifenhauser single-screw extruder. A significant feature of the die used was the presence of a spider consisting of a simple plate containing eighteen circular holes. Although this is commonly used as a means of removing the rotational memory of the melt in the screw, the melt will be subjected to elongational flow in its passage through each hole. This was important for the development of the particular microstructure observed.

Samples of the extrudate were etched with a strong oxidising acid known to attack the rubbery matrix (see [24]). Observations using scanning electron microscopy (SEM) indicated the fibrillar nature of the homopolymer in some of the samples, dependent on the volume fraction of polystyrene. For 10% PS, the blend showed no evidence of fibre formation, whereas for 50% PS almost perfect fibres were produced, as shown in Figure 9.6.

These observations were due to a combination of factors, namely a lowering of the surface energy of the dispersed homopolymer PS by the block copolymer and the formation of PS fibres by a process of elongational flow

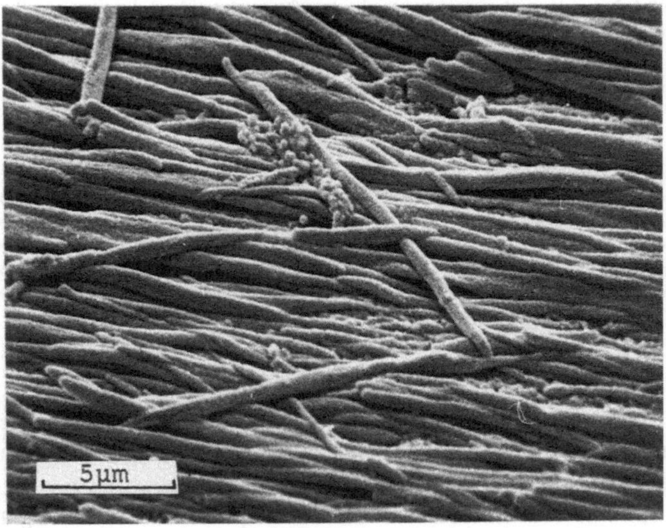

Figure 9.6 Scanning electron micrograph of an etched longitudinal section cut from a blend of Kraton 102 with 50% polystyrene, following the process of milling and extrusion. (From Ehtaiatkar *et al.* [23], by kind permission of Chapman and Hall.)

through the spider holes. When swollen in a selective agent for the polybutadiene phase of the S-B-S copolymer, it was found that dimensions of the blends changed in an anisotropic way; there being virtually no change parallel to the homopolymer PS fibrils and with all the swelling occurring laterally. The mechanical properties of the blends were also compatible with the presence of long continuous fibres of homopolymer PS embedded in the S-B-S matrix.

Overall, the microstructure and physical properties of these blends confirms the feasibility of producing fibre reinforced composite materials by the controlled extrusion of incompatible or partially incompatible blends. In these studies, the homopolymer PS fibres were formed, rather unexpectedly, by flow through the die assembly. Nevertheless, a principle has been established in this work and also in the work of Han and Kim [21], Tsebrenko *et al.* [20] and others, which has enabled the work to be taken further to develop a specific methodology for the fabrication of S-B-S/HDPE blends in which the HDPE has a high stiffness [25].

9.3 *In situ* fibre production in extruded blends

This section is concerned with the suitability of various polymer blends to form *stiff* fibres *in situ*. Full details of the findings are available in [25]. As a

Table 9.1 Viscosity and molecular weight data for grades of high density polyethylene. P denotes grade processed in power form.

Polymer grade	Melt flow index (MFI) (190 °C, 2.16 kg) (g/10 min)	$\bar{M}_n$	$\bar{M}_w$
Rigidex H020-54P	< 0.05	33 000	312 000
Rigidex H060-45P	0.15	26 300	265 000
Rigidex 9	1.10	6 060	126 600
Natene 60550AG(P)	7.00	14 600	72 800
Rigidex 140–60	13.00	13 350	67 800

result of its natural fibre forming ability, polyethylene has been chosen as the reinforcing phase and various grades of different molecular weight investigated, as listed in Table 9.1.

A matrix material was chosen which was capable of transferring an externally applied stress to the dispersed polyethylene phase, thereby aiding the drawing and extension process. The matrix was also chosen on the basis that it should be a melt at the processing temperature to act as a carrier for the dispersed phase, and have a low enough melting point to offer a wide window of operating temperatures. Immiscible systems in which the matrix had a melting point less than that of high density polyethylene were initially considered.

9.3.1 *Materials*

A block copolymer (Kraton 102) was chosen as the matrix material on the basis that it had a glass transition temperature (T_g) of approximately 90 °C and was viscous in its molten state. The three block copolymer consists of chains of the form polystyrene–polybutadiene–polystyrene (S-B-S). The molecular weight of the polystyrene (S) block is 10^4 and the polybutadiene (B) block 5.5×10^4, giving a volume fraction of polystyrene in the block copolymer of between 20% and 25%.

The structure and properties of this particular S-B-S copolymer have been extensively investigated over the past 10 to 20 years by Folkes and Keller [26], Fischer [27] and Keller *et al.* [28]. The microstructure of the copolymer consists of segregated phases of polystyrene within a matrix of polybutadiene. The polystyrene exists in the form of long cylinders, having a typical diameter of 15 nm. These cylinders may be very regularly arranged so as to form a macrolattice with spacings of the order of 30 to 40 nm. For the particular case of a cylindrical morphology the lattice is hexagonal, although in most specimens this ordering is usually only short range. If special care is taken, melt processed specimens can be fabricated such that most, if not all, of the material exhibits a uniformly oriented macrolattice or 'single-crystal'. Specimens such as this have been prepared by Folkes *et al.* [29] using extrusion. Indeed, the spatial ordering of the polystyrene phase can be such as to give discrete single-crystal low angle X-ray patterns [30].

From the point of view of the present investigation, the morphology of the block copolymer has at least one important implication. As a matrix for the blend, the copolymer melt will be 'structured' and this can lead to a high degree of molecular association between the matrix and the added homopolymer.

9.3.2 Formulation and compounding

The polyethylene grades described earlier were blended in ratios of 20/80, 40/60, 50/50, 60/40 and 80/20 wt % ratios with the S-B-S matrix material. The blends were initially mixed using a ribbon blade mixer and then compounded using a Gay's TS40-DVL corotating twin-screw extruder fitted with a die consisting of a simple plate containing six holes. The barrel temperature varied from 135 °C at the feed hopper to 180 °C at the die. The extrudate was fed directly to a cold water bath and kept free from tension at all times. Samples were then granulated in preparation for ram-extrusion or were left in extrudate form for drawing.

9.3.3 Microstructure of extruded S-B-S/HDPE blends

Micrographs of longitudinal sections in S-B-S/HDPE blends indicate that discrete polyethylene fibres exist in those grades where $M_w \leqslant 126\,600$. Figure 9.7 shows a section of the 60/40 S-B-S/60550AG blend in which the polyethylene exists as a series of macrofibres surrounded by the S-B-S matrix. Note that the dark areas represent the S-B-S and the light areas the polyethylene phase. Microfibres were observed after the same sample was immersed in immersion oil for approximately 15 minutes. The oil caused the rubbery phase to swell exposing fine polyethylene fibres with diameters of 1 μm.

Similar observations have been made in the blends containing the Rigidex 9 and Rigidex 140–60 polyethylene grades, although in the former case (higher M_w) the polyethylene did not form such well developed fibres. Those blends containing the highest molecular weight grades of Rigidex H020-54P and Rigidex H060-45P did not form fibres at the twin-screw extrusion stage. These results are very significant because the fibres were formed using a conventional twin-screw extrusion operation where no special die requirements were utilised. One explanation for this may be the lowering of surface energy of the dispersed homopolymer polyethylene by the block copolymer. The latter aids the formation of the fine fibrils by elongational or stretching flows, which may be present within the extruder barrel, for example, over the screw flights or even within the flights themselves. To confirm that the fibres were not formed in the die itself, the 60/40 S-B-S/60550AG blend was compounded using the

Figure 9.7 Longitudinal section of as-extruded 60/40 S-B-S/HDPE (60550AG) blend. The polyethylene (white area) is oriented in the extrusion direction.

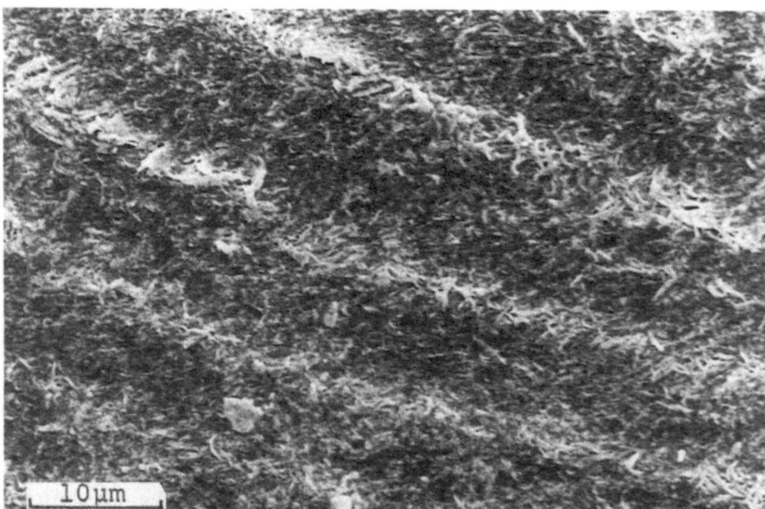

Figure 9.8 Transverse SEM micrograph of etched 60/40 S-B-S/HDPE(60550AG) blend (asextruded).

twin-screw extruder without the die in place. Analysis of the extruded blend indicated that polyethylene fibres still existed, confirming the above suggestion.

Another explanation for the ease with which the polyethylene fibres formed arises from an analysis of the matrix itself. As discussed earlier, the S-B-S block copolymer exists as a 'structured' melt, hence its high melt viscosity. During processing the polyethylene phase will be oriented, for example, over screw flights, but it is prevented from any molecular relaxation (recoil) by the structured S-B-S phase itself. That is, the PS rods within the S-B-S 'reinforce' the melt thereby preventing the polyethylene from obtaining its thermodynamically stable state and relaxing.

This observation is highlighted by Figure 9.8 which shows an etched SEM micrograph of a transverse section of the 60/40 S-B-S/60550AG blend. Some spatial ordering or layering of the HDPE phase is apparent in the sample and polyethylene fibres are clearly visible within the S-B-S matrix. Differential scanning calorimetry (DSC) and mechanical property measurements have indicated that no gross molecular orientation exists within the polyethylene phase. Further work [25] has indicated that highly oriented polyethylene fibres can be generated *in situ* by ram extrusion of the blends. This work is described in section 9.4.

9.4 Ram extrusion of S-B-S/HDPE blends

The S-B-S/HDPE blends were granulated and, using a capillary rheometer, then extruded at temperatures slightly above the melting point of the dispersed polyethylene phase. This temperature was chosen on the basis of work by Odell *et al.* [31] who produced oriented plugs of polyethylene using the die blockage technique.

Extrusion at these temperatures (e.g. 132 °C) was possible in the case of the S-B-S/HDPE blends because the matrix existed as a melt. The S-B-S matrix therefore acted as a carrier for the polyethylene particles, preventing die blockage, thereby allowing a continuous 'low temperature' extrusion process in which stiff polyethylene fibres could be produced.

The S-B-S/HDPE blends were initially extruded at 132 °C using a Davenport capillary rheometry. A range of shear rates was investigated to establish what conditions were required to form well-oriented polyethylene fibres in the blends. Results indicated that fibres only formed in those blends containing polyethylene grades with a molecular weight $\leqslant 126\,600$. This was because of the short residence times available to untangle the polymer chains in the high molecular weight materials. Examination of the S-B-S/60550AG blends indicated that the most perfectly formed fibres were produced in the 60/40 blend. Figure 9.9 shows the polyethylene fibre in a partially etched section of the blend. The apparent lack of orientation in this micrograph is due to the

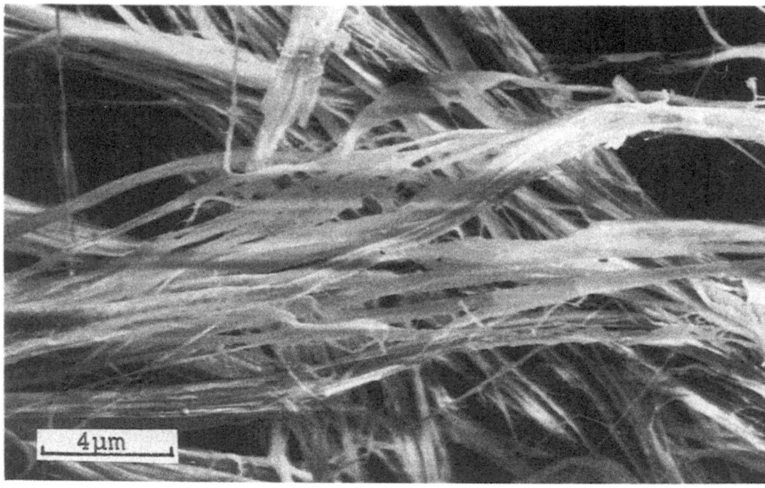

Figure 9.9 SEM micrograph of etched 60/40 S-B-S/HDPE(60550AG) blend (ram-extruded).

Figure 9.10 Optical micrograph under crossed polarisers of the ram extruded 60/40 S-B-S/HDPE blend.

collapse of the fibres after the partial removal of the S-B-S phase. As a result of the ease with which polyethylene fibres are formed in the 60/40 S-B-S/ 60550AG, further analysis has concentrated on this blend.

Ram extrusion of the blend at a temperature close to the melting point of the polyethylene phase results in a fairly stiff product which could not be cold drawn. Figure 9.10 shows a micrograph of the blend indicating that a definite core region exists in the centre of the sample, which is related to the extensional flow field, close to the axis of the capillary. Closer inspection has shown a 'shish-kebab' morphology in the core region and DSC endotherms show a double melting peak confirming the existence of this chain extended material [32–34].

9.4.1 *Effect of capillary die length on* in situ *fibre formation*

The effect of die length/diameter (L/D) ratio on polyethylene fibre formation is shown in Table 9.2. The results are summarised in terms of the DSC melting peak endotherm, Young's modulus (E) and birefringence (Δn) for the polyethylene phase. Young's modulus of the polyethylene has been calculated using a rule of mixtures analysis and full details of the technique are available in [25].

DSC endotherms for the blends indicate multiple melting peaks at 131 °C and 135 °C for all blends. Table 9.2 shows that the relative intensity of the two peaks changes depending on the processing conditions. The intensity of the second peak is proportional to the amount of higher melting point material present (i.e., chain extended material). Twin peaks result mainly because of a spread in crystal sizes for the polyethylene:extended chains (greater crystal size) giving rise to a higher melting point than chain folded ones. The increase in chain extended material is accompanied by an increase in Young's modulus and birefringence for the polyethylene phase.

The effect of a change in die L/D ratio is comparable to a change in the

Table 9.2 Effect of die L/D ratio on properties of polyethylene fibres in 60/40 S-B-S/ HDPE blends. Die diameter = 2 mm. Extrusion temperature = 132 °C. Shear rate = 30 s^{-1} (N/m^2 to psi: ÷ 6.89 × 10^3).

Die L/D	Die entrance pressure ($N/m^2 \times 10^7$)	Relative intensities of 131 °C and 135 °C peaks respectively	E (PE) GPa	Δn (PE) $\times 10^{-2}$
0.5	0.32	1.36:1.00	4.0	5.20
2.5	1.00	1.20:1.00	5.0	5.30
5.0	1.47	1.10:1.00	5.0	5.35
10.0	2.51	1.00:1.06	6.0	5.40
12.5	3.03	1.00:1.30	8.0	5.50
20.0	4.41	1.14:1.00	5.0	5.35

time-scale of the experiment. An analysis of the conditions required to obtain the desired effect indicates that in this process three time-scales are involved:

T_e = the time-scale of the experiment, which is primarily related to the residence time of the polymer in the capillary die
T_r = the relaxation time of the polymer, denoting the characteristic time in which the elastic stress relaxes and the chain orientation disappears
T_k = the time required for crystallisation to occur under the pertaining conditions of temperature and chain orientation

The balance between T_e and T_r will determine the orientation that is created in the molten polymer subjected to elongational flow. If $T_e \gg T_r$, the behaviour will be predominantly viscous with the formation of particles which, though deformed, will not contain an oriented structure. If $T_e \ll T_r$, the polymer will behave as a rubber elastic solid showing a high degree of orientation. However, if, in the latter case, crystallisation (T_k) occurs at a low orientation, the process will result in a product containing only moderately elongated particles of relatively low stiffness.

This balance is reflected in the results shown in Table 9.2 which indicates a critical L/D ratio of 12.5 for the experimental conditions used. Lower L/D ratios, that is, shorter residence times, do not allow sufficient time to deform the reinforcing phase. Conversely, long residence times allow relaxation/recoil of the polyethylene phase, lowering the degree of orientation and hence the melting point, Young's modulus and birefringence.

9.4.2 *Effect of ram extrusion temperature on* in situ *fibre formation*

The effect of ram extrusion temperature on the properties of the polyethylene fibres in a 60/40 S-B-S/HDPE blend is shown in Table 9.3. The results indicate

Table 9.3 Effect of ram extrusion temperature on properties of polyethylene fibres in 60/40 S-B-S/HDPE blends. Die diameter = 2 mm. Die length = 25 mm. Shear rate = $30\,s^{-1}$ (N/m^2 to psi: $\div 6.89 \times 10^3$).

Ram extrusion tempera-ture °C	Die entrance pressure ($N/m^2 \times 10^7$)	Relative intensities of 131 °C and 135 °C peaks respectively	E (PE) GPa	Δn (PE) $\times 10^{-2}$
127	3.58	1.50:1.00	5.0	5.35
128	3.44	1.50:1.00	5.0	5.30
129	3.39	1.10:1.00	6.0	5.35
130	3.34	1.00:1.36	8.0	5.50
131	3.10	137 °C only	9.5	5.60
132	3.03	1.00:1.30	8.0	5.40
133	2.96	5.80:1.00	4.0	5.00
134	2.86	131 °C only	4.0	4.90

a corresponding increase in Young's modulus and birefringence with increasing melting point, the most successful blend being the sample extruded at 131 °C.

The existence of a single melting peak at 137 °C for the sample extruded at 131 °C is even more significant when compared with the results of other workers such as Ward [35]. Starting with an almost identical polyethylene grade, Ward *et al.* have produced fibres with a Young's modulus of about 30 GPa having a melting point of 139 °C. Further work has shown that fibres with comparable properties can be generated *in situ*.

9.4.3 Effect of extrusion pressure on in situ *fibre formation*

A series of capillary dies was used to investigate the effect of ram extrusion pressure on the *in situ* fibre formation for the blends. Figure 9.11 shows the effect of die entrance pressure on the melting point of the polyethylene fibres produced during extrusion. Dies A to E show that a maximum melting point exists at a critical die entrance pressure. This confirms that a critical balance between the time required to orient, then elongate the molecules, and the time required for the molecules to relax is necessary to obtain a maximum enhancement in the degree of orientation and stiffness.

This implies that the ram extrusion process is even more commercially attractive than first thought because it involves the production of highly oriented polyethylene fibres using relatively low pressures. In fact, a typical sample produced during the trials had a fibre modulus of 13 GPa and a birefringence of 6.05×10^{-2}.

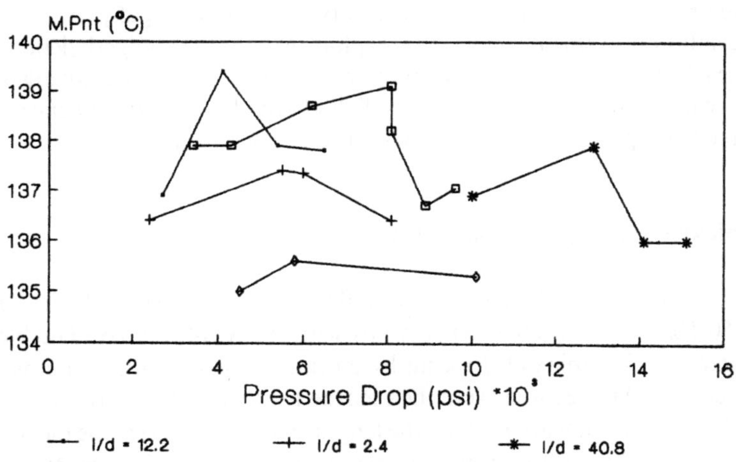

Figure 9.11 Effect of die entrance pressure on polyethylene melting point.

Figure 9.12 Etched transverse SEM section of 60/40 S-B-S/HDPE blend after ram extrusion.

An etched SEM micrograph of the sample is given in Figure 9.12 and shows fine polyethylene fibres embedded in the S-B-S matrix. The fibres have diameters of between 0.25 μm and 1 μm and are at least 1 mm in length, providing very high fibre aspect ratios indeed. One of the main disadvantages of the blends, however, is the poor dispersion of the polyethylene fibres. Some clumping and migration of the fibres was apparent and was due to the complex flow fields existing at the entrance to the capillary. This resulted in poorer than expected mechanical properties for the polyethylene fibres [25].

One solution to this problem is to exploit the fact that polyethylene fibres can be easily formed at the twin-screw extrusion stage. Uniaxial drawing of the blends will result in a 'well-dispersed' high modulus polyethylence fibre reinforced composite. This area of work is discussed in section 9.5.

9.5 Uniaxial drawing of polymer blends

The powdered polyethylene grade (60550AG) was blended in 20/80, 40/60, 50/50, 60/40 and 80/20 wt% ratios with the S-B-S matrix. The blends were prepared for uniaxial drawing by cutting the as-extruded lengths into 15 cm sections. The sections were then marked with a grid pattern, consisting of a series of lines 5 mm apart along the length. The draw ratio of the individual elements could then be monitored by measuring the ratio of the length of the drawn element to the undrawn element. A sample of 100% 60550AG polyethylene was also prepared and used as a control.

Details of the method used to calculate the Young's modulus of the polyethylene phase in the blends are given in [25]. Briefly, work by Arridge and Folkes [36] on ultrahigh modulus polyethylene has shown that very long samples with aspect ratios in excess of 100 are required before a true measure of Young's modulus can be obtained. In conventional tensometer measurements the load is applied to the sample via shear from the surface to interior points. For short samples, therefore, the load is applied to only a fraction of the sample cross-section, producing an abnormally large strain, hence underestimating the Young's modulus of the sample as a whole. As a consequence, drawn samples in excess of 100 mm in length were tested and a rule of mixtures analysis applied to the composite to determine the Young's modulus of the polyethylene phase:

$$E(\mathrm{PE}) = \frac{E(\mathrm{composite})}{v} - \frac{E(\mathrm{matrix})(l - v)}{v}$$

where v = volume fraction of polyethylene (PE).

The application of a tensile stress to the extruded 60/40 S-B-S/HDPE blend indicated that the sample formed a neck and resulted in a markedly stiffer product. In fact, there was a five-fold increase in Young's modulus (approximately 10 GPa) for the polyethylene phase in a sample drawn at a draw ratio of 11.0. Work by Capaccio and Ward [37] on high density polyethylene has shown that stricter control of the drawing process can lead to much higher draw ratios. In particular, it was shown that drawing at temperatures appreciably above ambient, but still below the polyethylene melting point, minimised the production of microvoids in the oriented samples on the one hand, and prevented the onset of 'flow drawing' on the other. This area of work, namely the achievement of higher draw ratios, has therefore been applied to the S-B-S/60550AG blends.

9.5.1 Hot drawing of polymer blends

The as-extruded blends were drawn at a series of temperatures between 20 °C and 105 °C, the latter being approximately 5 °C to 10 °C above the glass transition temperature (T_g) of the S-B-S phase. A series of draw rates (crosshead speeds) was also used, ranging from 20 mm/min to 500 mm/min. Note that the blends drawn at 500 mm/min broke prematurely and there was little success using draw temperatures above the T_g of the S-B-S phase because of the onset of flow drawing.

The effect of draw rate on draw ratio at a temperatrue of 90 °C for the various blends is shown in Figure 9.13. With the exception of the control sample (100% HDPE), all blends containing more than 40% polyethylene showed a corresponding increase in draw ratio with draw rate. This was most pronounced in the 50/50 S-B-S/6055AG blend which had a draw ratio of 20 at 400 mm/min. The poorer draw ratios obtained at the lower polyethylene

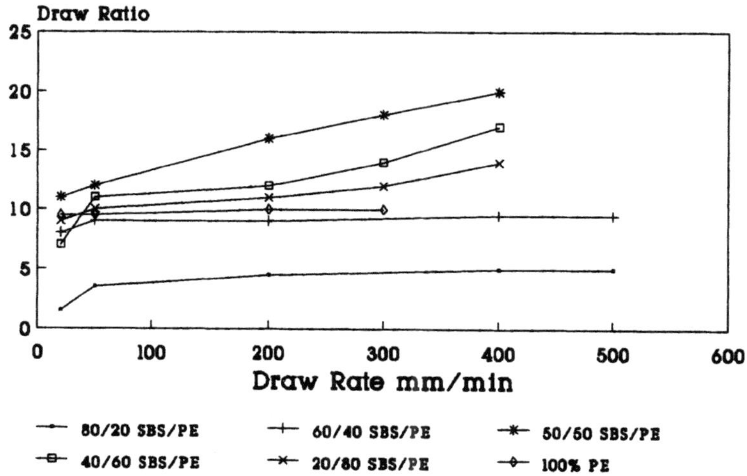

Figure 9.13 Effect of draw rate on draw ratio for different S-B-S/HDPE compositions. Draw temperature 90 °C.

weight fractions are due to the rubbery nature of the blends, which prevents the efficient stress transfer to the polyethylene phase. One explanation for the maximum draw ratio being at 50% polyethylene is that the S-B-S phase is effectively acting as a lubricant as well as an efficient stress transfer medium. As the polyethylene weight fraction is increased, the S-B-S phase is prevented from lubricating the fibres which are free to impinge on each other, thereby reducing the efficiency of the drawing operation.

Similar phenomena have been found to occur in draw ratio versus temperature data for the blends. Figure 9.14 shows that a critical draw temperature, which is dependent on the weight fraction of the polyethylene, exists for the blends. The data indicate that the draw ratio decreases initially with increasing temperature, and then rises again to some critical temperature (90 °C for the 50/50 S-B-S/60550AG blend). Above the T_g of the S-B-S phase the draw ratio falls to a minimum as a result of flow drawing. Below 50 °C the blend can be considered as a viscous medium which relies on the efficient transfer of stress between the S-B-S and the polyethylene phase. As the draw temperature is increased the S-B-S phase becomes less viscous and its efficiency as a stress transfer medium diminishes. This results in a slight decrease in draw ratio. Further increases in temperature allow the polyethylene phase to become more mobile during drawing resulting in a higher draw ratio.

The effect of draw rate and temperature on draw ratio, discussed above, was also reflected in the results for the polyethylene modulus. There was a linear relationship between draw ratio and polyethylene modulus for all the blends produced. The effect of polyethylene weight fraction on fibre modulus is shown in Figure 9.15. The 50/50 S-B-S/60550AG blend in Figure 9.15 was produced

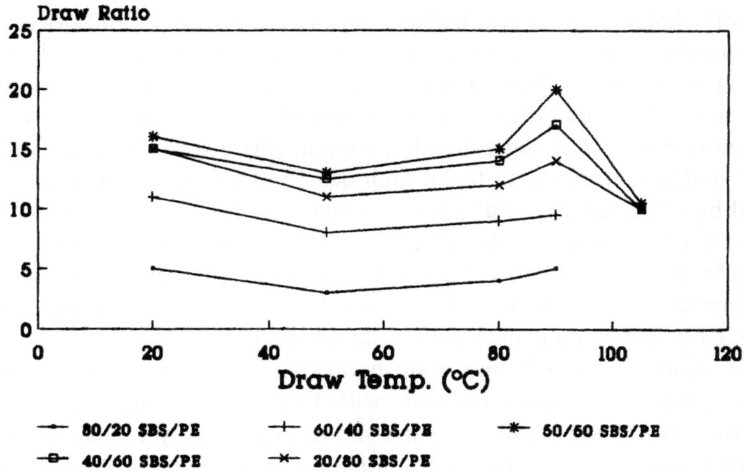

Figure 9.14 Effect of draw temperature on draw ratio for different S-B-S/HDPE compositions. Draw rate 400 mm/minute.

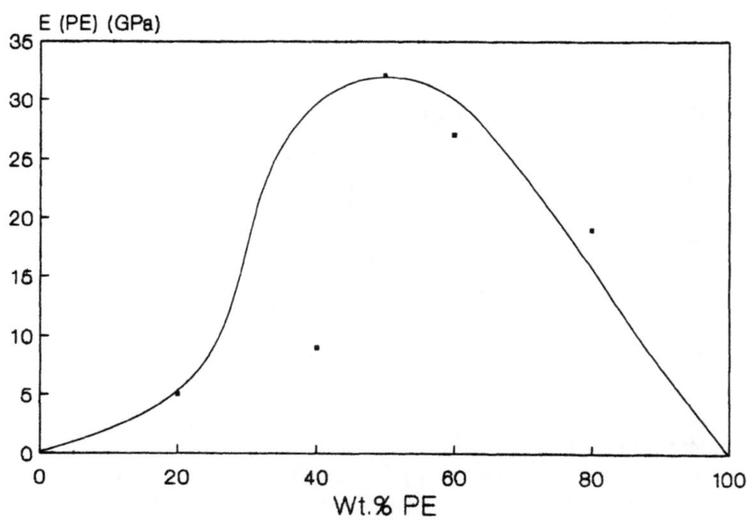

Figure 9.15 Effect of polyethylene weight fraction on fibre modulus for blends drawn at 400 mm/minute and 90 °C.

at a draw rate of 400 mm/min and a draw temperature of 90 °C. The result for this blend was very significant because polyethylene fibres with a Young's modulus of 32 GPa were formed *in situ*. This result is comparable with the results found by Capaccio and Ward [37], who achieved 40 GPa monofilaments at an equivalent draw ratio of 20 in samples of similar molecular weight.

The polyethylence fibrils also had similar melting behaviour and orientation birefringence to Ward's material. The crystalline melting point of the polyethylene fibres in the blends was 139.5 °C. However, the absence of multiple peaks in the DSC endotherm argued against the presence of a true extended chain morphology and suggested that the polyethylene structure was akin to that found in conventional drawn high density polyethylene. This was confirmed by estimates of crystal sizes from wide-angle X-ray diffraction photographs. Figure 9.16 shows the diffraction photograph for the most successful drawn 50/50 S-B-S/6055AG blend indicating a highly oriented product. The birefringence of the sample was measured as 6.15×10^{-2}, compared with 6.20×10^{-2} for a sample of Ward's material (see [37]).

This result is extremely important because it implies that very stiff fibres with a specific modulus comparable with E-glass can be generated *in situ*. Microstructure observations have also shown that the polyethylene fibrils are well dispersed in the composites. Examination has shown that the fibres have diameters of between 0.25 μm and 1 μm (Figure 9.17) with aspect ratios in excess of 1000.

Figure 9.16 X-ray diffraction photograph of 50/50 S-B-S/HDPE drawn blend. Draw rate 400 mm/minute at 90 °C.

Figure 9.17 Etched transverse SEM section of 50/50 S-B-S/HDPE drawn blend. Draw rate 400 mm/minute at 90 °C.

9.6 Conclusions and further aspects of *in situ* fibre forming technology

The term *in situ* fibre composite is widely applied to complex blend systems (via *in situ* crystallisation/polymerisation) or the expensive class of materials based on thermotropic liquid crystalline polymers. However, there is now a clear indication of the feasibility of producing stiff fibres *in situ*, by the controlled extrusion or drawing of a blend. This provides a basis for the development of a wide range of innovative composite materials, while at the same time avoiding many of the inherent difficulties associated with the processing of more traditional fibre reinforced thermoplastics.

The work described in sections 9.4 and 9.5 is applied to a relatively simple S-B-S/HDPE blend system. Using this work as a starting point, alternative blend systems based on a range of thermoplastic materials may be considered. Furthermore, the effect of molecular weight distribution on the fibre forming phase will play an important role. For example, the incorporation of a small amount of low molecular weight PE in the drawn blends may help the localised drawing process in the individual polyethylene particles. Alternatively, the incorporation of a high molecular weight fraction in the ram extruded blends will help form the chain extended shish-kebab structures described earlier. That is, the high molecular weight material can be used to form the chain extended core, the low molecular weight material nucleating along this chain extended region.

Both the ram extrusion and drawing routes may be incorporated into the production process itself. In the latter case, twin-screw extrusion of the blends may be followed by a controlled drawing process at the die exit to orient and extend the dispersed fibre-forming polymer within the matrix. The ram extrusion process can be conducted as a second stage to the twin-screw extrusion operation or conducted in-line with a ram extruder fitted to the end of a compounding facility. In either case, the final extrudate will consist of a matrix which is reinforced by parallel fibres of high stiffness and strength. Selection of a matrix material with a lower processing temperature than the melting point of the reinforcing fibres will allow the composite to be processed using compression and injection moulding technologies, without destroying the high draw ratio built into the fibre phase.

References

1. Paul, D.R. and Newman, S. (1978) *Polymer Blends*, Vols. 1 and 2, Academic Press, New York.
2. Han, C.D. (1981) *Multiphase Flow in Polymer Processing*, Academic Press, New York.
3. McHugh, A.J. (1975) *J. Appl. Polym. Sci.* **19**, 125.
4. Lyngaae-Jørgensen, J. (1985) in *Processing, Structure and Properties of Block Copolymers*, Folkes, M.J., ed., Applied Science Publishers, London, Chapter 3.
5. Porter, R.S. and Weeks, N. (1974) *J. Polym. Sci., Polym. Phys. Edn.* **12**, 635.
6. Capaccio, G. and Ward, I.M. (1973) *Nature Phys. Sci.* **243**, 143.
7. Ward, I.M. (1974) *J. Mat. Sci. (letters)* **9**, 1193.
8. Lemstra, P.J. and Smith, P. (1980) *J. Mat. Sci.* **15**, 505.
9. Van der Vegt, A.K. and Smit, P.P.A. (1967) *S.C.I. Monograph* **26**, 313.
10. Porter, R.S., Southern, J.H. and Weeks, N.E. (1975) *Polym. Eng. Sci.* **15**, 213.
11. Frank, F.C. (1970) *Proc. Roy. Soc. (A)* **319**, 127.
12. Capaccio, G. and Ward, I.M. (1974) *Polymer* **15**, 233.
13. Kiss, G., Kovacs, A.J. and Wittmann, J.C. (1981) *J. Appl. Polym. Sci.* **26**, 2665.
14. Joseph, J.R., Kardos, J.L. and Nielsen, L.E. (1968) *J. Appl. Polym. Sci.* **12**, 1151.
15. Wegner, G., Fischer, E.W. and Munoz-Escalona, A. (1975) *Makromol. Chem. Suppl.* **1**, 521.
16. Baughman, R.H., Gleiter, H. and Sendfield, N.J. (1975) *Polym. Sci., Polym. Phys. Edn.* **13**, 1871.
17. Baughman, R.H. (1981) US Patent 4 255 535.
18. Calundann, G.W. and Jaffe, M. (1982) *Proc. Robert A. Welsh Conferences on Chemical Research XXVI, Synthetic Polymers*, 247.
19. Kiss, G. (1987) *Polym. Eng. Sci.* **27**, 410.
20. Tsebrenko, M.V., Yudin, A.V., Ablazova, T.I. and Vinogradov, G. (1976) *Polymer* **17**, 831.
21. Han, C.D. and Kim, Y.M. (1974) *J. Appl. Polym. Sci.* **18**, 2589.
22. Folkes, M.J. and Reip, P.W. (1986) *Polymer* **27**, 377.
23. Ehtaiatkar, F., Folkes, M.J. and Steadman, S.C. (1989) *J. Mat. Sci.* **24**, 2808.
24. Bucknall, C.B. (1977) *Toughened Plastics*, Applied Science Publishers, London.
25. Steadman, S.C. (1989) The *in situ* production of polyethylene fibres from polymer blends, PhD Thesis, Brunel University.
26. Folkes, M.J. and Keller, A. (1973) in *Physics of Glassy Polymers*, Haward, R.N., ed., Applied Science Publishers, London.
27. Fischer, E. (1968) *J. Macromal. Sci. (Chem.)* **A2**, 1285.
28. Keller, A., Pedemonte, E. and Willmouth, F.M. (1970) *Nature* **255**, 538.
29. Folkes, M.J., Keller, A. and Scalisi, F.P. (1973) *Kolloid-Z, Polym.* **1**, 251.
30. Keller, A., Pedemonte, E. and Willmouth, F.M. (1970) *Kolloid-Z, Polym.* **238**, 385.
31. Odell, J.A., Grubb, D.T. and Keller, A. (1978) *Polymer* **19**, 617.
32. Grubb, D.T., Keller, A. and Odell, J.A. (1975) *J. Mat. Sci.* **10**, 1510.

33. Pennings, A.J. and Van der Mark, J.M.A.A. (1971) *Rheol. Acta.* **10**, 174.
34. Hellmuth, E. and Wunderlich, B. (1965) *J. Appl. Phys.* **36**, 3039.
35. Ward, I.M. (ed.) (1975) *Structure and Properties of Oriented Polymers*, Applied Science Publishers Ltd., London.
36. Arridge, R.G.C. and Folkes, M.J. (1976) *Polymer* **17**, 495.
37. Capaccio, G. and Ward, I.M. (1975) *Polym. Eng. Sci.* **15**, 219.

Index

MIX
Papier aus verantwortungsvollen Quellen
Paper from responsible sources
FSC® C105338

If you have any concerns about our products,
you can contact us on
ProductSafety@springernature.com

In case Publisher is established outside the EU,
the EU authorized representative is:
Springer Nature Customer Service Center GmbH
Europaplatz 3, 69115 Heidelberg, Germany

Printed by Libri Plureos GmbH
in Hamburg, Germany